MARTIN GARDNER

WHEELS, LIFE AND OTHER MATHEMATICAL AMUSEMENTS

W. H. Freeman and Company
New York

Library of Congress Cataloging in Publication Data
Gardner, Martin, 1914–
  Wheels, life, and other mathematical amusements.
  Includes bibliographies and index.
  1. Mathematical recreations. I. Title.
QA95.G333 1983   793.7'4   83-11592
ISBN 0-7167-1588-0
ISBN 0-7167-1589-9 (pbk.)

Life configurations courtesy of R. William Gosper of Symbolics.

Jacket photograph by Jennifer Walsh

PRINTED IN THE UNITED STATES OF AMERICA

34567890 VB 5432108987

For Ronald L. Graham

who juggles numbers
and other mathematical objects
as elegantly
as he juggles balls and clubs, and twirls himself
on the trampoline

CONTENTS

INTRODUCTION

> "There remains one more game."
> "What is it?"
> "Ennui," I said. "The easiest of all. No rules, no boards, no equipment."
> "What is Ennui?" Amanda asked.
> "Ennui is the absence of games."
>
> —Donald Barthelme, *Guilty Pleasures.*

Unfortunately, as recent studies of education in this country have made clear, one of the chief characteristics of mathematical classes, especially on the lower levels of public education, is ennui. Some teachers may be poorly trained in mathematics and others not trained at all. If mathematics bores them, can you blame their students for being bored?

Like science, mathematics is a kind of game that we play with the universe. The best mathematicians and the best teachers of mathematics obviously are those who both understand the rules of the game, and who relish the excitement of playing it. Raymond Smullyan, who has enormous zest for the games of philosophy and mathematics, once taught an elementary course in geometry. In his delightful book *5000 B.C. and Other Philosophical Fantasies* (1983) he tells how he once introduced his students to the Pythagorean theorem:

> I drew a right triangle on the board with squares on the hypotenuse and legs and said, "Obviously, the square on the hypotenuse has a larger area than either of the other two squares. Now suppose these three squares were made of beaten gold, and you were offered either the one large square or the two small squares. Which would you choose?"
>
> Interestingly enough, about half the class opted for the one large square and half for the two small ones. A lively argument began. Both groups were equally amazed when told that it would make no difference.

It is this sense of surprise that all great mathematicians feel, and all great teachers of mathematics are able to communicate. I know of no better way to do this, especially for beginning students, than by way of games, puzzles, paradoxes, magic tricks, and all the other curious paraphernalia of "recreational mathematics."

"Puzzles and games provide a rich source of example problems useful for illustrating and testing problem-solving methods," wrote Nils Nilsson in his widely used textbook *Problem-Solving Methods in Artificial Intelligence.* He quotes Marvin Minsky: "It is not that the games and mathematical problems are chosen because they are clear and simple; rather it is that *they give us, for the smallest initial structures, the greatest complexity,* so that one can engage some really formidable situations after a relatively minimal diversion into programming."

Nilsson and Minsky had in mind the value of recreational mathematics in learning how to solve problems by computers, but its value in learning how to solve problems by hand is just as great. In this book, the tenth collection of the Mathematical Games columns that I wrote for *Scientific American,* you will find an assortment of mathematical recreations of every variety. The last three chapters (the third was written especially for this volume) deal with John H. Conway's fantastic game of Life, the full wonders of which are still being explored.

The two previously published articles on Life, in which I had the privilege of introducing this game for the first time, aroused more interest among computer buffs around the world than any other columns I have written. Now that Life software is becoming available for home-computer screens, there has been a renewed interest in this remarkable recreation. Although Life rules are incredibly simple, the complexity of its structure is so awesome that no one can experiment with its "life forms" without being overwhelmed by a sense of the infinite range and depth and mystery of mathematical structure. Few have expressed this emotion more colorfully than the British-American mathematician James J. Sylvester:

> Mathematics is not a book confined within a cover and bound between brazen clasps, whose contents it needs only patience to ransack; it is not a mine, whose treasures may take long to reduce into possession, but which fill only a limited number of veins and lodes; it is not a soil, whose fertility can be exhausted by the yield of successive harvests; it is not a continent or an ocean, whose area can be mapped out and its contour defined: it is limitless as that space which it finds too

narrow for its aspirations; its possibilities are as infinite as the worlds which are forever crowding in and multiplying upon the astronomer's gaze; it is as incapable of being restricted within assigned boundaries or being reduced to definitions of permanent validity, as the consciousness, the life, which seems to slumber in each monad, in every atom of matter, in each leaf and bud and cell, and is forever ready to burst forth into new forms of vegetable and animal existence.

Martin Gardner

1

WHEELS

> The miraculous paradox of smooth round
> objects conquering space by simply tumbling
> over and over, instead of laboriously lifting
> heavy limbs in order to progress, must have
> given young mankind a most salutary shock.

—VLADIMIR NABOKOV, *Speak, Memory*

Things would be very different without the wheel. Transportation aside, if we consider wheels as simple machines—pulleys, gears, gyroscopes and so on—it is hard to imagine any advanced society without them. H. G. Wells, in *The War of the Worlds*, describes a Martian civilization far ahead of ours but using no wheels in its intricate machinery. Wells may have intended this to be a put-on; one can easily understand how the American Indian could have missed discovering the wheel, but a society capable of sending spaceships from Mars to the earth?

Until recently the wheel was believed to have originated in Mesopotamia. Pictures of wheeled Mesopotamian carts date back to 3000 B.C. and actual remains of massive disk wheels have been unearthed that date back to 2700 B.C. Since World War II, however, Russian archaeologists have found pottery models of wheeled carts in the Caucasus that suggest the wheel may have originated in southern Russia even earlier than it did in Mesopotamia. There could have been two or more independent inventions. Or it may have spread by cultural diffusion as John Updike describes it in a stanza of his poem, *Wheel:*

> The Eskimos had never heard
> Of centripetal force when Byrd
> Bicycled up onto a floe
> And told them, "This how white man go."

It seems surprising that evolution never hit on the wheel as a means for making animals go, but on second thought one realizes how difficult it would be for biological mechanisms to make wheeled feet rotate. Perhaps the tumbleweed is the closest nature ever came to wheeled transport. (On the other hand, the Dutch artist Maurits C. Escher designed a creature capable of curling itself into a wheel and rolling along at high speeds. Who can be sure such creatures have not evolved on other planets?) There may also be submicroscopic swivel devices inside the cells of living bodies on the earth, designed to unwind and rewind double-helix strands of DNA, but their existence is still conjectural.

A rolling wheel has many paradoxical properties. It is easy to see that points near its top have a much faster ground speed than points near its bottom. Maximum speed is reached by a point on the rim when it is exactly at the top, minimum speed (zero) when the point touches the ground. On flanged train wheels whose rims extend slightly below a track, there is even a short segment in which a point on the rim moves backward. G. K. Chesterton, in an essay on wheels in his book *Alarms and Discursions,* likens the wheel to a healthy society in having "a part that perpetually leaps helplessly at the sky; and a part that perpetually bows down its head into the dust." He reminds his readers, in a characteristically Chestertonian remark, that "one cannot have a Revolution without revolving."

The most subtle of all wheel paradoxes is surprisingly little known, considering that it was first mentioned in the *Mechanica,* a Greek work attributed to Aristotle but more likely written by a later disciple. "Aristotle's wheel," as the paradox is called, is the subject of a large literature to which such eminent mathematicians as Galileo, Descartes, Fermat and many others contributed. As the large wheel in Figure 1 rolls from *A* to *B,* the rim of the small wheel rolls along a parallel line from *C* to *D.* (If the two lines are actual tracks, the double wheel obviously cannot roll smoothly along both. It either rolls on the upper track while the large wheel continuously slides backward on the lower track, or it rolls on the lower track while the small wheel slips forward on the upper track. This is not, however, the heart of the paradox.) Assume that the bottom wheel rolls without slipping from *A* to *B.* At every instant that a unique

Figure 1

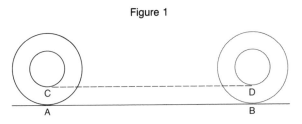

Aristotle's wheel paradox

point on the rim of the large wheel touches line *AB*, a unique point on the small wheel is in contact with line *CD*. In other words, all points on the small circle can be put into one-to-one correspondence with all points on the large circle. No points on either circle are left out. This seems to prove that the two circumferences have equal lengths.

Aristotle's wheel is closely related to Zeno's well-known paradoxes of motion, and it is no less deep. Modern mathematicians are not puzzled by it because they know that the number of points on any segment of a curve is what Georg Cantor called *c*, the transfinite number that represents the "power of the continuum." All points on a one-inch segment can be put in one-to-one correspondence with all points on a line a million miles long as well as on a line of infinite length. Moreover, it is not hard to prove that there are aleph-one points within a square or cube of any size, or within an infinite Euclidian space having any finite number of dimensions. Of course, mathematicians before Cantor were not familiar with the peculiar properties of transfinite numbers, and it is amusing to read their fumbling attempts to resolve the wheel paradox.

Galileo's approach was to consider what happens when the two wheels are replaced by regular polygons such as squares [*see Figure* 2]. After the large square has made a complete turn along *AB*, the sides of the small square have coincided with *CD*

Figure 2

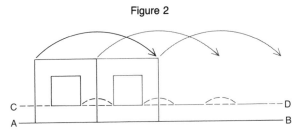

Galileo's approach to the wheel paradox

in four segments separated by three jumped spaces. If the wheels are pentagons, the small pentagon will jump four spaces on each rotation, and so on for higher-order polygons. As the number of sides increases, the gaps also increase in number but decrease in length. When the limit is reached—the circle with an infinite number of sides—the gaps will be infinite in number but each will be infinitely short. These Galilean gaps are none other than the mystifying "infinitesimals" that later so muddied the early development of calculus.

And now we are in a quagmire. If the gaps made by the small wheel are infinitely short, why should their sum cause the wheel to slide a finite distance as the large wheel rolls smoothly along its track? Readers interested in how later mathematicians replied to Galileo, and argued with one another, will find the details in the articles listed in this chapter's bibliography.

As a wheel travels a straight line, any point on its circumference generates the familiar cycloid curve. When a wheel rolls on the inside of a circle, points on its circumference generate curves called hypocycloids. When it rolls on the outside of a circle, points on the circumference generate epicycloids. Let R/r be the ratio of the radii, R for the large circle, r for the small. If R/r is irrational, a point a on the rolling circle, once in contact with point b on the fixed circle, will never touch b again even though the wheel rolls forever. The curve generated by a will have an aleph-null infinity of cusps. If R/r is rational, a and b will touch again after a finite number of revolutions. If R/r is integral, a returns to b after exactly one revolution.

Consider hypocycloids traced by a circle of radius r as it rolls inside a larger circle of radius R. When R/r is 2, 3, 4, . . . , points a and b touch again after one revolution and the curve will have R/r cusps. For example, a three-cusped deltoid results when R/r equals 3 [*see Figure* 3, *left*]. The same deltoid is produced when R/r is 3/2; that is, when the rolling circle's radius

Figure 3

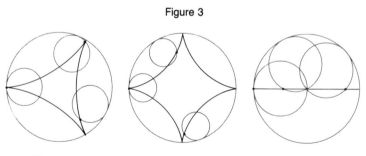

The deltoid   The astroid   "Two-cusped" hypocycloid

is two-thirds that of the fixed circle. All line segments tangent to the deltoid, with ends on the curve, have the same length. A four-cusped astroid is generated when R/r equals 4 or 4/3 [*see Figure* 3, *middle*]. The two ratios apply to all higher-order hypocycloids of this type: when R/r is either n or $n/(n-1)$, the rolling circle produces an n-cusped curve.

There is a surprising result when R/r equals 2 [*see Figure* 3, *right*]. The hypocycloid degenerates into a straight line coinciding with a diameter of the larger circle. Its two ends may be regarded as degenerate cusps. Can you guess the shape of the region swept over by a given diameter of the smaller circle? It is a region bounded by an astroid. This is the same as saying that the astroid is the envelope of a line segment that rotates while it keeps its ends on two perpendicular axes, as shown in Figure 4.

Figure 4

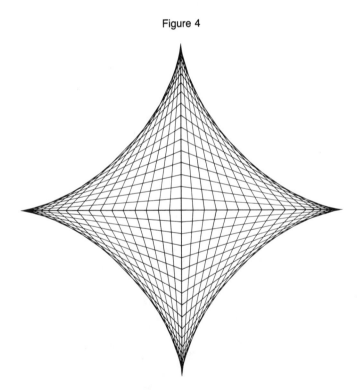

Astroid drawn as the envelope of a moving line segment

The simplest case of an epicycloid traced by a point on the rim of a wheel rolling outside another circle is seen when the two circles are equal. The result is a heart-shaped curve called

Figure 5

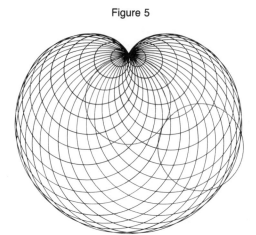

The cardioid

the cardioid [*see Figure* 5]. All chords drawn through its cusp have the same length. The cardioid in the illustration was drawn by dividing the fixed circle into 32 equal arcs and then drawing a set of circles whose centers are on this fixed circle and that pass through other points on the same circle. The figure can be shaded to produce a dazzling Op-art pattern [*see Figure* 6]. (Both pictures are from Hermann von Baravalle, *Geometrie als Sprache der Formen*, Stuttgart, 1963.)

Figure 6

Op-art cardioid

The cardioid is also generated by a point on the circumference of a circle that rolls twice around a fixed circle inside it that is half as large in diameter. This fact underlies a problem that was incorrectly answered in *The American Mathematical Monthly* for December, 1959 (Problem E 1362) but correctly answered in the March 1960 issue of the same journal. Imagine a girl whose bare waist is a perfect circle. Rolling around her waist, while she remains motionless, is a hula hoop with a diameter twice that of her waist. When a point on the hoop, touching the girl's navel, first returns to her navel, how far has that point traveled? Since the point traces a cardioid, this is equivalent to asking for the cardioid's length. It is not hard to show that it is four times the diameter of the hoop or eight times the diameter of the girl's waist.

When a rolling circle is half the diameter of a fixed circle that it touches externally, the epicycloid is the two-cusped nephroid (meaning kidney-shaped) that is shown in Figure 7. The drawing both shows the rolling circle and demonstrates a method of constructing the nephroid as the envelope of circles whose centers are on the fixed circle and that are tangent to the vertical central axis. As before, the curve can also be generated by rolling a circle around a smaller circle inside it; in this case, when R/r is 3/2. This is the same ratio as that which produces a deltoid, but now it is the larger circle that does the rolling.

Figure 7

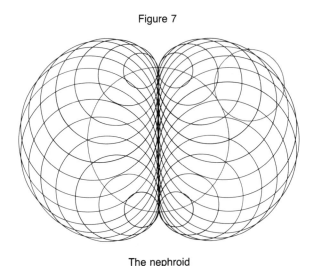

The nephroid

The cardioid and the nephroid are both caustics, curves enveloped by reflected light rays. The cardioid appears when the rays originate at a point on the circumference and are reflected by the circumference. The nephroid is produced by parallel rays crossing the circle, or from rays originating at the cusp of a cardioid and reflected by the cardioid. The cusped curve that one often sees on the surface of tea or coffee in a cup, when slanting light falls across the liquid from a window or other light source far to one side, is a good approximation of a nephroid cusp. Pleasant approximations are also frequently seen on photographs that appear in girlie magazines.

There are varied and perplexing problems that involve noncircular "wheels." For example, suppose a square wheel rolls without slipping on a track that is a series of equal arcs, convex sides up. What kind of curve must each arc be to prevent the center of the wheel from moving up and down? (In other words, the wheel's center must travel a straight horizontal path.) The curve is a familiar one and, amazingly, the same curve applies to similar tracks for wheels that are regular polygons with any number of sides. The answer will be disclosed in the answer section at the end of this chapter.

And can any reader solve this new riddle from Stephen Barr: What type of conveyance has eight wheels, carries only one person and never pollutes the atmosphere?

ANSWERS

The main problem was to describe the track that allows a square wheel to roll along it so that its center travels a straight horizontal line. The track is a series of catenary arcs. This applies to all wheels that are regular polygons. (If a wheel is an irregular convex polygon, the track must have arcs that are differently shaped catenaries, one for each side of the wheel.) If the wheel turns with a constant speed, its horizontal speed will vary. For details of the proof I must refer readers to "Rockers and Rollers," by Gerson B. Robison, in *Mathematics Magazine* for January, 1960, pages 139–144, and the solution to Problem *E*1668 in *The American Mathematical Monthly* for January, 1965, pages 82–83.

The riddle's answer is a pair of roller skates.

ADDENDUM

When I said that a point at the top of a wheel moves faster with respect to the ground than any other point on the wheel, I could have added that it moves exactly twice as fast as the center of the wheel. A. J. Knisely called attention to this in a short article, "The Rolling Wheel," in *Scientific American*, July 1891, and described a simple way of demonstrating it with a spool of thread.

George Lenfestey wrote to say that although he enjoyed my column on the wheel, it ruined his day:

> The trouble is, I've been sitting here wasting the better part of the afternoon imagining that gorgeous blue-eyed blond girl of yours in the hip huggers and halter top, twirling that hula hoop around her perfectly-formed golden middle. Please try to be more considerate in the future.

In my column I spoke of how difficult it would be for evolution to introduce a wheel into living organisms. A few years later, to my amazement, I read in *Scientific American* ("How Bacteria Swim," by Howard C. Berg) about the discovery that bacteria rotate their flagella like tiny propellers! In his *Oz* books L. Frank Baum introduced the "Wheelers" who have four wheels instead of four feet, and a bird called the "Ork" that flies by means of a propeller on its tail. These creatures are of course as imaginary as Escher's rolling animal or the fabled "hoop snake" that is said to bite its tail and roll like a hoop. However, there are spiders in Africa that actually escape from predators by curling into a ball and rolling down a hill.

BIBLIOGRAPHY

"The Cardioid." Robert C. Yates. *The Mathematics Teacher*, Vol. 52, January 1959, pages 10–14.
A Book of Curves. E. H. Lockwood. Cambridge University Press, 1961.

On Aristotle's Wheel:

"Aristotle's Wheel: Notes on the History of the Paradox." Israel Drabkin. *Osiris*, Vol. 9, 1950, pages 162–198.
"The Wheel of Aristotle and French Consideration of Galileo's Arguments." Pierre Costabel. *The Mathematics Teacher*, Vol. 61, May 1968, pages 527–534.
"The Wheel of Aristotle." David Ballew. *The Mathematics Teacher*, Vol. 65, October, 1972, pages 507–509.

Origin of the Wheel:

"The Beginnings of Wheeled Transport." Stuart Piggott. *Scientific American*, July 1968, pages 82–90.

2

DIOPHANTINE ANALYSIS

AND FERMAT'S LAST THEOREM

> The methods of Diophantus
> May cease to enchant us
> After a life spent trying to gear 'em
> To Fermat's Last Theorem.
>
> —J. A. LINDON, *A Clerihew*

An old chestnut, common in puzzle books of the late 19th century (when prices of farm animals were much lower than today), goes like this. A farmer spent $100 to buy 100 animals of three different kinds. Each cow cost $10, each pig $3 and each sheep 50 cents. Assuming that he bought at least one cow, one pig and one sheep, how many of each animal did the farmer buy?

At first glance this looks like a problem in elementary algebra, but the would-be solver quickly discovers that he has written a pair of simultaneous equations with three unknowns, each of which must have a value that is a positive integer. Finding integral solutions for equations is today called Diophantine analysis. In earlier centuries such analysis allowed integral fractions as values, but now it is usually restricted to whole numbers, including zero and negative integers. Of course in problems such as the one I have cited the values must be positive integers. Diophantine problems abound in puzzle literature. The well-known problem of the monkey and the coconuts, and the ancient task of finding right-angle triangles with integral sides, are among the classic instances of Diophantine problems.

The term "Diophantine" derives from Diophantus of Alexandria. He was a prominent Greek mathematician of his time, but to this day no one knows in what century he lived. Most authorities place him in the third century A.D. Nothing is known about him except some meager facts contained in a rhymed problem that appeared in a later collection of Greek puzzles. The verse has been quoted so often and its algebraic solution is so trivial, that I shall not repeat it here. If its facts are correct, we know that Diophantus had a son who died in his middle years and that Diophantus lived to the age of 84. About half of his major work, *Arithmetica*, has survived. Because many of its problems call for a solution in whole numbers, the term Diophantine became the name for such analysis. Diophantus made no attempt at a systematic theory, and almost nothing is known about Diophantine analysis by earlier mathematicians.

Today Diophantine analysis is a vast, complex branch of number theory with an enormous literature. There is a complete theory only for linear equations. No general method is known (it may not even exist) for solving equations with powers of 2 or higher. Even the simplest nonlinear Diophantine equation may be fantastically difficult to analyze. It may have no solution, an infinity of solutions or any finite number. Scores of such equations, so simple a child can understand them, have resisted all attempts either to find a solution or to prove none is possible.

The simplest nontrivial Diophantine equation has the linear form $ax + by = c$, where x and y are two unknowns and a, b and c are given integers. Let us see how such an equation underlies the puzzle in the opening paragraph. Letting x be the number of cows, y the number of pigs and z the number of sheep, we can write two equations:

$$10x + 3y + z/2 = 100$$

$$x + y + z = 100$$

To get rid of the fraction, multiply the first equation by 2. From this result, $20x + 6y + z = 200$, subtract the second equation. This eliminates z, leaving $19x + 5y = 100$. How do we find integral values for x and y? There are many ways, but I shall give only an old algorithm that utilizes continued fractions and that applies to all equations of this form.

Put the term with the smallest coefficient on the left: $5y = 100 - 19x$. Dividing both sides by 5 gives $y = (100 - 19x)/5$. We next divide 100 and $19x$ by 5, putting the remainders (if any) over 5 to form a terminal fraction. In this way the equation is transformed to $y = 20 - 3x - 4x/5$.

It is obvious that if x and y are positive integers (as they must be), x must have a positive value that will make $4x/5$ an integer. Clearly x must be a multiple of 5. The lowest such multiple is 5 itself. This gives y a value of 1 and z (going back to either of the two original equations) a value of 94. We have found a solution: 5 cows, 1 pig, 94 sheep. Are there other solutions? If negative integers are allowed, there are an infinite number, but here we cannot allow negative animals. When x is given a value of 10, or any higher multiple of 5, y becomes negative. The problem therefore has only one solution.

In this easy example the first integral fraction obtained, $4x/5$, does not contain a y term. For equations of the same form but with larger coefficients, the procedure just described must often be repeated many times. The terminal fraction is made equal to a new unknown integer, say a, the term with the lowest coefficient is put on the left, and the procedure is repeated to obtain a new terminal fraction. Eventually you are sure to end with a fraction that has only one unknown and is simple enough so that you can see what values the unknown must have to make the fraction integral. By working backward through whatever series of equations has been necessary, the original problem is solved.

For an example of an equation similar to the one just analyzed that has *no* solution, assume that cows cost $5, pigs $2 and sheep 50 cents. The two equations are handled exactly as before. The first is doubled to eliminate the fraction and the second is subtracted, producing the Diophantine equation $9x + 3y = 100$. Using the procedure of continued fractions, you end with $y = 33 - 3x - 1/3$, which shows that if x is integral, y cannot be. In this case, however, we can tell at once that $9x + 3y = 100$ has no solution by applying the following old theorem. If the coefficients of x and y have a common factor that is not a factor of the number on the right, the equation is unsolvable in integers. In this case 9 and 3 have 3 as a common divisor, but 3 is not a factor of 100. It is easy to see why the theorem holds. If the two terms on the left are each a multiple of n, so will their sum be; therefore the term on the right also must be a multiple of n. An even simpler instance would be $4x + 8y = 101$. The left side of the equality obviously must be an even integer, so that it cannot equal the odd number on the right. It is also good to remember that if all three given numbers do have a common factor, the equation can immediately be reduced by dividing all terms by the common divisor.

As an example of a variant of the basic problem that has a finite number (more than one) of positive-integer solutions, consider the case in which cows cost $4, pigs $2 and sheep a third of a dollar. As before, the farmer spends $100 on 100 animals, buying at least one of each. How many of each does he buy?

Many geometric problems are solved by finding integral solutions for Diophantine equations. In the chapter on triangles in my *Mathematical Circus* I gave two classic examples: Finding integer solutions for a problem involving two crossed-ladders, and for a problem concerning the location of a spot inside an equilateral triangle. Among the many geometrical Diophantine problems that are still unsolved, one of the most difficult and notorious is known as the problem of the "integral brick" or "rational cuboid." The "brick" is a rectangular parallelepiped. There are seven unknowns: The brick's three edges, its three face diagonals, and the space diagonal that goes from one corner through the brick's center to the opposite corner [*see Figure 8*]. Can a brick exist for which all seven variables have integer values?

Figure 8

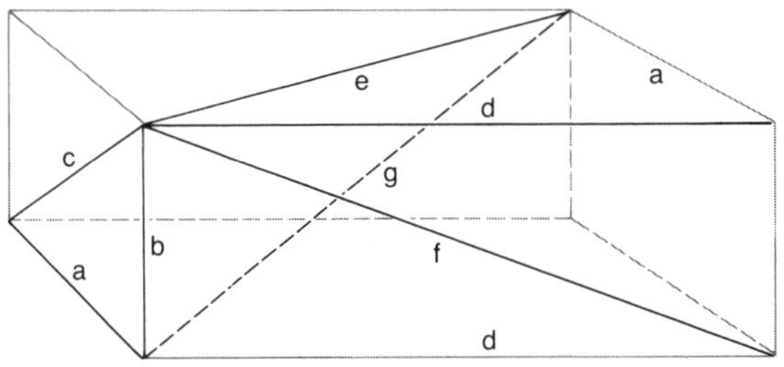

The integral brick, an unsolved Diophantine problem

The problem is equivalent to finding integer solutions for the seven unknowns in the following set of equations:

$$a^2 + b^2 = c^2$$
$$a^2 + d^2 = e^2$$
$$b^2 + d^2 = f^2$$
$$b^2 + e^2 = g^2$$

The problem has not been shown to be impossible, nor has it been solved. John Leech, a British mathematician, has been searching for a solution, and I am indebted to him for the following information. The smallest brick with integral edges and face diagonals (only the space diagonal is nonintegral) has edges of 44, 117 and 240. This was known by Leonhard Euler to be the minimum solution. If all values are integral except a face diagonal, the smallest brick has edges of 104, 153 and 672, a result also known to Euler. (The brick's space diagonal is 697). The third case, where only an edge is nonintegral, has not, as far as Leech knows, been considered before. It too has solutions, but the numbers are, as Leech puts it, "hideous." He suspects that the smallest such brick may be one with edges of 7,800, 18,720, and the irrational square root of 211, 773, 121. Of course the brick's volume is also irrational.

A much easier geometric problem, which I took from a puzzle book by L. H. Longley-Cook, is illustrated in Figure 9. A rectangle (the term includes the square) is drawn on graph paper as shown and its border cells are shaded. In this case the shaded cells do not equal the unshaded cells of the interior rectangle. Is it possible to draw a rectangle of proportions such that the border—one cell wide—contains the same number of cells as there are within the border? If so, the task is to find *all* such solutions. The Diophantine equation that is involved can be solved easily by a factoring trick, which I shall explain in the answer section.

Figure 9

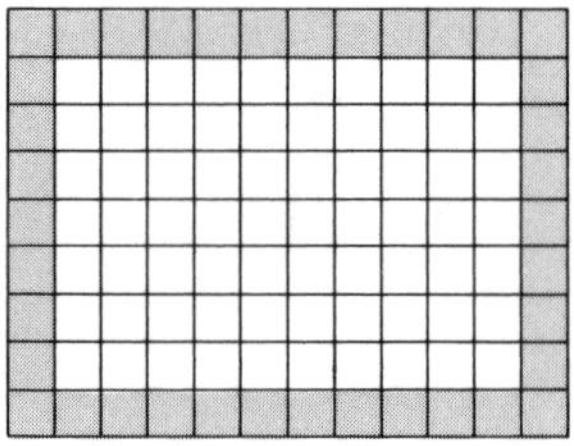

A simple Diophantine problem

In ancient times the most famous Diophantine problem, posed by Archimedes, became known as the "cattle problem." It involves eight unknowns, but the integral solutions are so huge (the smallest value contains more than 200,000 digits) that it was not solved until 1965 when a computer managed to do it. The interested reader will find a good discussion of the cattle problem in Eric Temple Bell's *The Last Problem,* and the final solution, by H. C. Williams and others, in the journal *Mathematics of Computation* (see bibliography).

The greatest of all Diophantine problems, still far from solved, is the "last theorem" of Pierre de Fermat, the 17th-century French amateur number theorist. (He was a jurist by profession.) Every mathematician knows how Fermat, reading Diophantus' *Arithmetica,* added a note in Latin to the eighth problem of the second book, where an integral solution is asked for $x^2 + y^2 = a^2$. Fermat wrote that such an equation had no solution in integers for powers greater than 2. (When the power is 2, the solution is called a "Pythagorean triple" and there are endless numbers of solutions.) In brief, Fermat asserted that $x^n + y^n = a^n$ has no solution in integers if n is a positive integer greater than 2. "I have discovered a truly marvelous demonstration," Fermat concluded his note, "which this margin is too narrow to contain."

To this day no one knows if Fermat really had such a proof. Because the greatest mathematicians since Fermat have failed to find a proof, the consensus is that Fermat was mistaken. Lingering doubts arise from the fact that Fermat always *did* have a proof whenever he said he did. For example, consider the Diophantine equation $y^3 = x^2 + 2$. It is easy to find by trial and error that it has the solutions $3^3 = 5^2 + 2$ and $3^3 = (-5)^2 + 2$. To prove, however, that there are no other integral solutions, Bell writes in *Men of Mathematics,* "requires more innate intellectual capacity . . . than it does to grasp the theory of relativity." Fermat said he had such a proof although he did not publish it. "This time he was not guessing," Bell continues. "The problem is hard; he asserted that he had a proof; a proof was later found." Fermat did publish a relatively elementary proof that $x^4 + y^4 = a^4$ has no solution, and later mathematicians proved the impossibility of the more difficult $x^3 + y^3 = a^3$. The cases of $n = 5$ and $n = 7$ were settled early in the 19th century.

It can be shown that Fermat's last theorem is true if it holds for all prime exponents greater than 2. By 1978 the theorem had been proved for all exponents less than 125,000, so if there is a counterexample it will involve numbers with more than a million digits. Proving the theorem continues to be the deepest unsolved problem in Diophantine theory. Some mathematicians believe it may be true but unprovable, now that Kurt Gödel has shown, in his famous undecidability proof, that arithmetic contains theorems that cannot be established inside the deductive system of arithmetic. (If Fermat's last theorem is Gödelian-undecidable, it would have to be true, because if it were false, it would be decidable by a single counterexample.)

I earnestly ask readers *not* to send me proofs. I am not competent to evaluate them. Ferdinand Lindemann, the first to prove (in 1882) that pi is transcendent, once published a long proof of Fermat's last theorem that turned out to have its fatal mistake right at the beginning. Dozens of other fallacious proofs have been published by leading mathematicians. When David Hilbert was asked why he never tackled the problem, his reply was: "Before beginning I should put in three years of intensive study, and I haven't that much time to squander on a probable failure."

The mathematics departments of many large universities return all proofs of Fermat's last theorem with a form letter stating that the paper will be evaluated only after an advance payment of a specified fee. Edmund Landau, a German mathematician, used a form letter that read: "Dear Sir/Madam: Your proof of Fermat's last theorem has been received. The first mistake is on page _____ , line _____ ." Landau would then assign the filling in of the blanks to a graduate student.

Donald E. Knuth whimsically asks for a proof of Fermat's last theorem as the last exercise at the end of his preface to the first volume of his series *The Art of Computer Programming* (1968). His answer states that someone who read a preliminary draft of the book reported that he had a truly remarkable proof but that the margin of the page was too small to contain it.

Euler failed to prove Fermat's last theorem, but he made a more general conjecture that, if it is true, would include the truth of Fermat's last theorem as a special case. Euler suggested that no nth power greater than 2 can be the integral sum of fewer than n nth powers. As we have seen, it has long been known that the conjecture holds when n is 3, for this is merely Fermat's last theorem with powers of 3. It is not yet known whether or not $x^4 + y^4 + z^4 = a^4$ has a solution.

In 1966, about two centuries after Euler made his guess, a counterexample was published. Leon J. Lander and Thomas R. Parkin, with the help of a computer program, showed that Euler's conjecture fails for $n = 5$. The counterexample with the lowest coefficients is:

$$27^5 + 84^5 + 110^5 + 133^5 = 144^5.$$

This result suggests that if there are intelligent creatures living somewhere in a space of five dimensions, their puzzle books surely contain the following problem. What is the smallest hypercube of five dimensions that can be built with unit hypercubes such that the same number of unit hypercubes will form four smaller hypercubes, with no unit hypercubes left over? The answer is a cube of $144 \times 144 \times 144 \times 144 \times 144$ units.

ANSWERS

1. The problem about the farmer and the animals reduces to the Diophantine equation $11x + 5y = 200$. Applying the method of continued fractions, three solutions in positive integers can be found:

Cows	Pigs	Sheep
5	29	66
10	18	72
15	7	78

2. L. H. Longley-Cook, in *Fun with Brain Puzzlers* (Fawcett, 1965), Problem 87, solves the rectangle problem as follows. Let x and y be the sides of the large rectangle. The total number of cells it contains is xy. The border, one cell wide, contains $2x + 2y - 4$ cells. Since we are told that the border contains $xy/2$ cells, we can write the equation:

$$xy/2 = 2x + 2y - 4.$$

Double both sides and rearrange the terms:

$$xy - 4x - 4y = -8.$$

Add 16 to each side:

$$xy - 4x - 4y + 16 = 8.$$

The left side can be factored:

$$(x - 4)(y - 4) = 8.$$

It is clear that $(x-4)$ and $(y-4)$ must be positive integral factors of 8. The only pairs of such factors are 8, 1 and 4, 2. They provide two solutions: $x = 12$, $y = 5$, and $x = 8$, $y = 6$.

The problem is closely related to integral-sided right triangles. The width of the border is an integer only when the diagonal of the large rectangle cuts it into two such "Pythagorean triangles."

If we generalize the problem to allow nonintegral solutions for borders of any uniform width, keeping only the proviso that the area of the border be equal to the area of the rectangle within it, there is an unusually simple formula for the width of the border. (I am indebted to S. L. Porter for it.) Merely add two adjacent sides of the border, subtract the diagonal of the large rectangle and divide the result by four. This procedure gives the width of the border.

Several readers generalized this problem to three dimensions, seeking integral edges for a brick composed of unit cubes equal to the number of unit cubes required to cover it on all sides with a one-unit layer of cubes. Daniel Sleator used a computer to find the complete solution, a total of 20 bricks. The smallest uncovered brick has edges of 6, 8, 10; the largest, 3, 11, 130. This confirms a guess made by M. H. Greenblatt in *Mathematical Entertainments* (Crowell, 1965), page 11, that the problem has "about" 20 solutions.

ADDENDUM

One of the most famous of all unsolved problems in Diophantine theory, the so-called Hilbert's tenth problem, was brilliantly solved in 1970 by Yu. V. Matijasevic, a 22-year-old graduate student at the University of Leningrad. In 1900 the great German mathematician David Hilbert compiled a list of 23 outstanding unsolved problems that he hoped would be solved during the twentieth century. Problem 10 was to find a general algorithm that would decide whether any given polynomial Diophantine equation, with integer coefficients, has a solution in integers.

Matijasevic proved that there is no such algorithm. In other words, he "solved" Hilbert's tenth problem by proving it had no solution. The Fibonacci number sequence plays a key role in his proof.

For details see "Hilbert's Tenth Problem," by Martin Davis and Reuben Hersh in *Scientific American,* November 1973, pages 84–91, and "Hilbert's Tenth Problem is Unsolvable," by Martin Davis in *The American Mathematical Monthly,* Volume 80, March 1973, pages 233–269.

BIBLIOGRAPHY

Diophantus of Alexandria. Sir Thomas L. Heath. Dover, 1964.

Diophantine Equations. Louis Joel Mordell. Academic Press, 1969.

On Euler's conjecture:

"Counterexample to Euler's Conjecture on Sums of Like Powers." L. J. Lander and T. R. Parkin. *Bulletin of the American Mathematical Society,* Vol. 72, 1966, page 1079.

"A Counterexample to Euler's Sum of Powers Conjecture." L. J. Lander and T. R. Parkin. *Mathematics of Computation,* Vol. 21, January 1967, pages 101–103.

"A Survey of Equal Sums of Like Powers." L. J. Lander, T. R. Parkin, and J. L. Selfridge. *Mathematics of Computation*, Vol. 21, July 1967, pages 446–459.

On Fermat's Last Theorem:

Fermat's Last Theorem. A. Church. Long Island University Press, 1937.

Murder by Mathematics. Hector Hawton. London: Ward Lock and Co., 1948.

"The Devil and Simon Flagg." Arthur Porges in *Fantasia Mathematica*, edited by Clifton Fadiman. Simon and Schuster, 1958.

The Last Problem. Eric Temple Bell. Simon and Schuster, 1961.

"Fermat's Last Theorem." Harold M. Edwards. *Scientific American*, October 1978, pages 104–121.

"13 Lectures on Fermat's Last Theorem." Paulo Ribenboim. Springer-Verlag, 1979.

On the Monkey and the Coconuts:

"The Monkey and the Coconuts." Martin Gardner. *The Second Scientific American Book of Mathematical Puzzles & Diversions*, Chapter 9. Simon and Schuster, 1961.

On Pythagorean Triples:

"The Pythagorean Theorem." Martin Gardner. *The Sixth Book of Mathematical Games from Scientific American*, Chapter 16. W. H. Freeman, 1971.

On Archimedes' Cattle Problem:

The Last Problem. Eric Temple Bell. Simon and Schuster, 1961, pages 151–157.

"Solution of the Cattle Problem of Archimedes." H. C. Williams, R. A. German, and C. R. Zarnke. *Mathematics of Computation*, Vol. 19, October 1965, page 671. See also comments by Daniel Shanks, *ibid.*, pages 686–687.

"A Solution to Archimedes Cattle Problem." Harry L. Nelson. *Journal of Recreational Mathematics*, Vol. 13, 1980–81, pages 162–176. See also Nelson's note in Vol. 14, 1981–82, page 126.

On the Integral Brick:

"On the Integral Cuboid." W. G. Spohn. *The American Mathematical Monthly*, Vol. 79, January 1972, pages 57–59.

"On the Derived Cuboid." W. G. Spohn. *Canadian Mathematical Bulletin*, Vol. 17, 1974, pages 575–577.

"The Rational Cuboid Revisited." John Leech. *The American Mathematical Monthly*, Vol. 84, August 1977, pages 518–533. See Vol. 85, June 1978, page 472, for corrections.

Unsolved Problems in Number Theory. Richard K. Guy (Springer-Verlag, 1981), pages 97–101.

3

THE KNOTTED MOLECULE

AND OTHER PROBLEMS

1. THE KNOTTED MOLECULE

Enormously long chainlike molecules (long in relation to their breadth) have been discovered in living organisms. The question has arisen: Can such molecules have knotted forms? Max Delbrück of the California Institute of Technology, who received a Nobel prize in 1960, proposed the following idealized problem:

Assume that a chain of atoms, its ends joined to form a closed space curve, consists of rigid, straight-line segments each one unit long. At every node where two such "links" meet, a 90-degree angle is formed. At each end of each link, therefore, the next link may have one of four different orientations. The entire closed chain could be traced along the edges of a cubical lattice, with the proviso that at each node the joined links form a right angle [*see Figure* 10]. At no point is the chain allowed to touch or intersect itself; that is, two and only two links meet at every node.

Figure 10

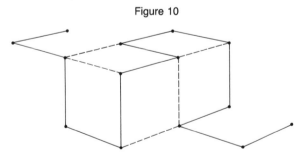

Example of a 13-link chain

What is the shortest chain of this type that is tied in a single overhand (trefoil) knot? In the answer I shall reproduce the shortest chain Delbrück has found. It has not been proved minimal; perhaps a reader will discover a shorter one. (I wish to thank John McKay for calling this problem to my attention.)

2. PIED NUMBERS

The old problem of expressing integers with four 4's (discussed in a *Scientific American* column reprinted as Chapter 5 of *The Incredible Dr. Matrix*) has been given many variations. In an intriguing new variant proposed by Fitch Cheney one is allowed to use only pi and symbols for addition, subtraction, multiplication, division, square root and the "round-down function." In the last operation, indicated by brackets, one takes the greatest integer that is equal to or less than the value enclosed by the brackets. Parentheses also may be used, as in algebra, but no other symbols are allowed. Each symbol and pi may be repeated as often as necessary, but the desideratum is to use as few pi symbols as possible. For example, 1 can be written [√π] and 3 even more simply as [π].

The reader is invited to do his best to express the integers from 1 through 20 according to these rules, and to compare them with the best Cheney was able to achieve.

3. THE FIVE CONGRUENT POLYGONS

L. Vosburgh Lyons contributed a fiendish dissection problem to a magic magazine in 1969 [*see Figure* 11]. The polygon [*at left in illustration*] can be dissected into four congruent polygons [*at right*]. Can the reader discover the only way in which the same polygon can be cut into *five* congruent polygons?

Figure 11

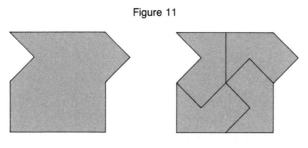

L. Vosburgh Lyons' dissection problem

4. STARTING A CHESS GAME

A full set of 32 chessmen is placed on a chessboard, one piece to a cell. A "move" consists in transferring a piece from the cell it is on to any empty cell. (This has nothing to do with chess moves.) Gilbert W. Kessler, a mathematics teacher in a Brooklyn high school, thought of the following unusual problem: How can you place the 32 pieces so that a maximum number of transfer moves are required to arrange the pieces in the correct starting position for a game of chess?

It is not specified which side of the board is the black side, but the playing sides must, as in regulation chess, be sides with a white square in the bottom right corner, and of course the queen must go on a square matching her own color. One is tempted at first to think that the maximum is 33 moves, but the problem is trickier than that.

5. THE TWENTY BANK DEPOSITS

A Texas oilman who was an amateur number theorist opened a new bank account by depositing a certain integral number of dollars, which we shall call x. His second deposit, y, also was an integral number of dollars. Thereafter each deposit was the sum of the two previous deposits. (In other words, his deposits formed a generalized Fibonacci series.) His 20th deposit was exactly a million dollars. What are the values of x and y, his first two deposits? (I am indebted to Leonard A. Monzert of West Newton, Mass., for sending the problem of which this is a version.)

The problem reduces to a Diophantine equation that is somewhat tedious to solve, but a delightful shortcut using the golden ratio becomes available if I add that x and y are the two positive integers that begin the longest possible generalized Fibonacci chain ending in a term of 1,000,000.

6. THE FIRST BLACK ACE

A deck of 52 playing cards is shuffled and placed face down on the table. Then, one at a time, the cards are dealt face up from the top. If you were asked to bet in advance on the distance from the top of the first black ace to be dealt, what position (first, second, third, . . .) would you pick so that if the game were repeated many times, you would maximize your chance in the long run of guessing correctly?

7. A DODECAHEDRON-QUINTOMINO PUZZLE

John Horton Conway defines a "quintomino" as a regular pentagon whose edges (or triangular segments) are colored with
five different colors, one color to an edge. Not counting rotations and reflections as being different, there are 12 distinct
quintominoes. Letting 1, 2, 3, 4, 5 represent the five colors, the
12 quintominoes can be symbolized as follows:

A.	12345	G.	13245
B.	12354	H.	13254
C.	12435	J.	13425
D.	12453	K.	13524
E.	12534	L.	14235
F.	12543	M.	14325

The numbers indicate the cyclic order of colors going either
clockwise or counterclockwise around the pentagon [*see Figure
12, left*]. In 1958 Conway asked himself if it was possible to
color the edges of a regular dodecahedron [*Figure 12, middle*]
in such a way that each of the 12 quintominoes would appear
on one of the solid's 12 pentagonal faces. He found that it was
indeed possible. Can readers find a way to do it?

Figure 12

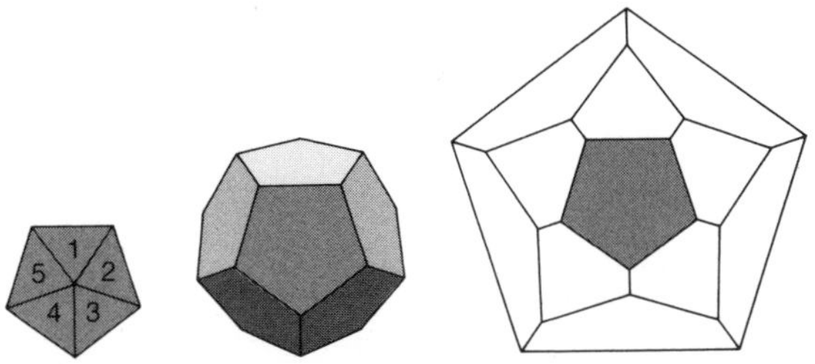

The A quintomino   The dodecahedron   Schlegel diagram of dodecahedron

Those who like to make mechanical puzzles can construct a
cardboard model of a dodecahedron with small magnets glued
to the inside of each face. The quintominoes can be cut from
metal and colored on both sides (identical colors opposite each
other) so that any piece can be "reflected" by turning it over.
The magnets, of course, serve to hold the quintominoes on the
faces of the solid while one works on the puzzle. The problem
is to place the 12 pieces in such a way that the colors match
across every edge.

Without such a model, the Schlegel diagram of a dodecahedron [*Figure 12, right*] can be used. This is simply the distorted

skeleton of the solid, with its back face stretched to become the figure's outside border. The edges are to be labeled (or colored) so that each pentagon (including the one delineated by the pentagonal perimeter) is a different quintomino.

8. SCRAMBLED QUOTATION

Letters in the sentence "Roses are red, violets are blue" are scrambled by the following procedure. The words are written one below the other and flush at the left:

ROSES
ARE
RED
VIOLETS
ARE
BLUE

The columns are taken from left to right and their letters from the top down, skipping all blank spaces, to produce this ordering:

RARVABOREIRLSEDOEUELESETS.

The task is to find the line of poetry that, when scrambled by this procedure, becomes

TINFLABTTULAHSORIOOASAWEIKOKNARGEKEDYE-ASTE.

Walter Penney of Greenbelt, Md., contributed this novel word problem to the February 1970 issue of *Word Ways: The Journal of Recreational Linguistics*. That lively quarterly is currently being published privately by A. Ross Eckler, Spring Valley Road, Morristown, N.J. 07960.

9. THE BLANK COLUMN

A secretary, eager to try out a new typewriter, thought of a sentence shorter than one typed line, set the controls for the two margins and then, starting at the left and near the top of a sheet of paper, proceeded to type the sentence repeatedly. She typed the sentence exactly the same way each time, with a period at the end followed by the usual two spaces. She did not, however, hyphenate any words at the end of a line: When she saw that the next word (including whatever punctuation marks may have followed it) would not fit the remaining space on a line, she shifted to the next line. Each line, therefore, started flush at the left with a word of her sentence. She finished the page after typing 50 single-spaced lines.

Without experimenting on a typewriter, answer this question: Is there sure to be at least one perfectly straight column of blank spaces on the sheet, between the margins, running all the way from top to bottom? (T. Robert Scott originated this problem, which was sent to me by his friend W. Lloyd Milligan of Columbia, S.C.)

10. THE CHILD WITH THE WART

A: "What are the ages, in years only, of your three children?"
B: "The product of their ages is 36."
A: "Not enough information."
B: "The sum of their ages equals your house number."
A: "Still not enough information."
B: "My oldest child—and he's at least a year older than either of the others—has a wart on his left thumb."
A: "That's enough, thank you. Their ages are. . . ."
Complete *A*'s sentence. (Mel Stover of Winnipeg was the first of several readers to send this problem, the origin of which I do not know.)

ANSWERS

1. The shortest knotted chain known that meets all the conditions specified has 36 links [*see Figure* 13]. It is reproduced from Max Delbrück's paper, "Knotting Problems in Biology,"

Figure 13

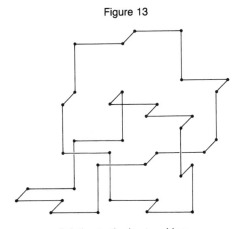

Solution to the knot problem

in *Mathematical Problems in the Biological Sciences: Proceedings of Symposia in Applied Mathematics,* Vol. 14, 1962, pages 55–68.

If links are not required to be at right angles to their adjacent links, knots of 24 links are possible.

2. Figure 14 gives Fitch Cheney's answers to the problem of expressing the integers 1 through 20 by using pi, as few times as possible, and the symbols specified. He was able to express all integers from 1 through 100 without using more than four pi's in each expression.

Figure 14

$$1 = [\sqrt{\pi}]$$
$$2 = [\sqrt{\pi}\ \sqrt{\pi}]$$
$$3 = [\pi]$$
$$4 = [\pi + \sqrt{\pi}]$$
$$5 = [\pi\ \sqrt{\pi}]$$
$$6 = [\pi + \pi]$$
$$7 = [\pi^{\sqrt{\pi}}]$$
$$8 = [(\pi \times \pi) - \sqrt{\pi}]$$
$$9 = [\pi \times \pi]$$
$$10 = [\pi \times \pi] + [\sqrt{\pi}]$$

$$11 = [(\pi \times \pi) + \sqrt{\pi}]$$
$$12 = [\pi \times \pi] + [\pi]$$
$$13 = [(\pi \times \pi) + \pi]$$
$$14 = [(\pi \times \pi) + \pi + \sqrt{\pi}]$$
$$15 = [\pi \times \pi] + [\pi + \pi]$$
$$16 = [(\pi \times \pi) + \pi + \pi]$$
$$17 = [\pi \times \pi \times \sqrt{\pi}]$$
$$18 = [\pi \times \pi] + [\pi \times \pi]$$
$$19 = [(\pi \times \pi) + (\pi \times \pi)]$$
$$20 = [\pi\sqrt{\pi}]\ [\pi + \sqrt{\pi}]$$

How the first 20 integers can be "pied"

Hundreds of readers improved on Cheney's answers. Here are some typical ways of shortening six of the expressions:

$$14 = [\ [\pi] \times (\pi + \sqrt{\pi})]$$
$$15 = [\pi] \times [\pi\sqrt{\pi}]$$
$$16 = [\pi\sqrt{\pi} \times [\pi]]$$
$$18 = [\pi] \times [\pi + \pi]$$
$$19 = [\pi(\pi + \pi)]$$
$$20 = [\pi^{\pi}/\sqrt{\pi}]\ \text{or}\ [(\pi\sqrt{\pi})^{\sqrt{\pi}}]$$

These improvements reduce the total number of pi's to 50. John W. Gosling was the first of many readers to achieve 50, but it is only fair to add that the problem did not specifically allow exponention and that many who wrote earlier than Gosling would probably have achieved 50 had they used exponents for integers 7 and 20. (Without exponents, 7 requires three pi's and 20 requires four.) Numerous readers lowered the number of pi's below 50 by adding other symbols, such as the factorial sign or the "unary negative operator," which has the effect of rounding up instead of down. Bernard Wilde and Carl Thune, Mark T. Longley-Cook, V. E. Hoggatt, Jr., Robert L. Caswell

and others conjectured that by using nested radical signs to reduce a divisor, any positive integer can be expressed with three
pi's.

Cheney and John Leech each pointed out that if $-[-\pi]$ is
interpreted in a standard way to mean 4, then further reduc-
tions are possible:

$$2 = -[-\sqrt{\pi}]$$
$$4 = -[-\pi]$$
$$8 = -[-\pi]-[-\pi]$$
$$10 = -[-\pi \times \pi]$$
$$11 = [(-[-\pi]^{\sqrt{\pi}}]$$
$$12 = [-\pi \times (-\pi)]$$
$$13 = -[\pi \times (-\pi)]$$
$$16 = [-\pi] \times [-\pi]$$

3. The large polygon in Figure 15 can be cut into five congruent polygons as shown. The method obviously enables one
to dissect the polygon into any desired number of congruent
shapes. L. Vosburgh Lyons first published this in *The Pallbearers Review* for July, 1969, page 268.

Figure 15

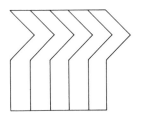

Solution to the dissection problem

4. The 32 chess pieces can be placed so that 36 "moves" are
needed to transfer the pieces to a correct starting position with
black at the top and white at the bottom [*see Figure* 16].

It was stated in the problem that it was not necessary for
black to be at the top. However, if the final position is black at
the bottom, then 37 moves are required to produce a starting
pattern with the queens on the right color. If it is required that
black be at the top, then a standard starting position, with
white at the bottom, requires 38 moves to effect the change.

5. The Texas oilman's bank deposit problem reduces to the
Diophantine equation $2{,}584x + 4{,}181y = 1{,}000{,}000$. It can be
solved by Diophantine techniques such as the continued-frac-

Figure 16

A solution to the chess problem

tion method explained in Chapter 2. The first two deposits are $154 and $144.

The shortcut, given that *x* and *y* start the longest possible Fibonacci chain terminating in 1,000,000, rests on the fact that the longer a generalized Fibonacci series continues, the closer the ratio of two adjacent terms approaches the golden ratio. To find the longest generalized Fibonacci chain that ends with a given number, place the number over *x* and let it equal the golden ratio. In this case the equation is

$$\frac{1,000,000}{x} = \frac{1 + \sqrt{5}}{2}$$

Solve for *x* and change the result to the nearest integer. It is 618,034. Because no other integer, when related to 1,000,000, gives a closer approximation of the golden ratio, 618,034 is the next-to-last term of the longest possible chain of positive integers in a generalized Fibonacci series ending in 1,000,000. One can now easily work backward along the chain to the first two terms. (This method is explained in Litton Industries' *Problematical Recreations*, edited by Angela Dunn, Booklet 10, Problem 41.)

6. Contrary to most people's intuition, your best bet is that the top card is a black ace.

The situation can be grasped easily by considering simpler cases. In a packet of three cards, including the two black aces and, say, a king, there are three equally probable orderings: *AAK, AKA, KAA.* It is obvious that the probability of the first ace's being on top is 2/3 as against 1/3 that it is the second card. For a full deck of 52 cards the probability of the top card's being the first black ace is 51/1,326, the probability that the first black ace is second is 50/1,326, that it is third is 49/1,326, and so on down to a probability of 1/1,326 that it is the 51st card. (It cannot, of course, be the last card.)

In general, in a deck of n cards (n being equal to or greater than 2) the probability that the first of two black aces is on top is $n-1$ over the sum of the integers from 1 through $n-1$. The probability that the first black ace is on top in a packet of four cards, for instance, is 1/2.

The problem is given by A. E. Lawrence in "Playing with Probability," in *The Mathematical Gazette*, Vol. 53, December 1969, pages 347–354. As David L. Silverman has noticed, by symmetry the most likely position for the *second* black ace is on the bottom. The probability for each position of the second black ace decreases through the same values as before but in reverse order from the last card (51/1,326) to the second from the top (1/1,326).

Several readers pointed out that the problem of the first black ace is a special case of a problem discussed in *Probability with Statistical Applications*, by Frederick Mosteller, Robert E. K. Rourke and George B. Thomas, Jr. (Addison-Wesley, 1961). Mosteller likes to call it the "needle in the haystack" problem and give it in the practical form of a manufacturer who has, say, four high-precision widgets randomly mixed in his stock with 200 low-precision ones. An order comes for one high-precision widget. Is it cheaper to search his stock or to tool up and make a new one? His decision depends on how likely he is to find one near the beginning of a search. In the case of the 52-card deck there is a better-than-even chance that an ace will be among the first nine cards at the top of a shuffled deck or— what amounts to the same thing—among the first nine cards picked at random without replacement.

7. The three essentially distinct solutions of the dodecahedron-quintomino puzzle are shown on Schlegel diagrams in Figure 17. They were first published by John Horton Conway, the inventor of the problem, in the British mathematical journal *Eureka* for October, 1959, page 22. Each solution has a mirror reflection, of course, and colors can be interchanged without altering the basic pattern. The letters correspond to those assigned previously to the 12 quintominoes. The letter outside each diagram denotes the quintomino on the solid's back face, represented by the diagram's perimeter.

Figure 17

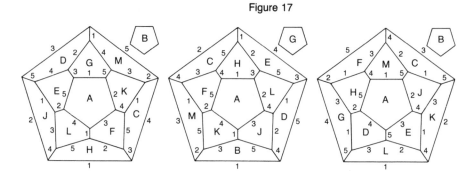

Three basic solutions to the dodecahedron-quintomino problem

Conway found empirically that whenever the edges of 11 faces were correctly labeled, the 12th face was automatically labeled to correspond with the remaining quintomino. He did not prove that this must always be true.

Because the regular dodecahedron is the "dual" of the regular icosahedron, the problem is equivalent to coloring the edges of the regular icosahedron so that at its 12 vertexes the color permutations correspond to the permutations of colors on the 12 quintominoes.

In 1972 a white plastic version of Conway's puzzle was on sale in the United States under the name "Enigma." Patterns of black dots were used instead of colors.

8. The scrambled quotation is "There is no frigate like a book/To take us lands away." It is the first two lines of a poem by Emily Dickinson.

9. There is certain to be at least one column of blank spaces on that typed page. Assume that the sentence is n spaces long, including the first space following the final period. This chain of n spaces will begin each typed line, although the chain may be cyclically permuted, beginning with different words in different lines. Consequently the first n spaces of every line will be followed by a blank space.

10. Only the following eight triplets have a product of 36: 1, 1, 36; 1, 2, 18; 1, 3, 12; 1, 4, 9; 1, 6, 6; 2, 2, 9; 2, 3, 6, and 3, 4, 3. Speaker A certainly knew his own house number. He would therefore be able to guess the correct triplet when he was told it had a sum equal to his house number—unless the sum was 13, because only two triplets have identical sums, $1 + 6 + 6$ and $2 + 2 + 9$, both of which equal 13. As soon as A was told that B had an oldest child he eliminated 1, 6, 6, leaving 2, 2, 9 as the ages of B's three children.

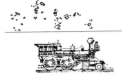

4

ALEPHS AND SUPERTASKS

Points
Have no parts or joints.
How then can they combine
To form a line?

—J. A. LINDON

Every finite set of n elements has 2^n subsets if one includes the original set and the null, or empty, set. For example, a set of three elements, ABC, has $2^3 = 8$ subsets: ABC, AB, BC, AC, A, B, C, and the null set. As the philosopher Charles Sanders Peirce once observed (*Collected Papers* 4. 181), the null set "has obvious logical peculiarities." You can't make any false statement about its members because it has no members. Put another way, if you say anything logically contradictory about its members, you state a truth, because the solution set for the contradictory statement is the null set. Put colloquially, you are saying something true about nothing.

In modern set theory it is convenient to think of the null set as an "existing set" even though it has no members. It can also be said to have 2^n subsets because $2^0 = 1$, and the null set has one subset, namely itself. And it is a subset of every set. If set A is included in set B, it means that every member of set A is a member of set B. Therefore, if the null set is to be treated as a legitimate set, all its members (namely none) must be in set B. To prove it by contradiction, assume the null set is *not* included in set B. Then there must be at least one member of the null set that is not a member of B, but this is impossible because the null set has no members.

The n elements of any finite set obviously cannot be put into one-to-one correspondence with its subsets because there are always more than n subsets. Is this also true of infinite sets? The answer is yes, and the general proof is one of the most beautiful in all set theory.

It is an indirect proof, a *reductio ad absurdum*. Assume that all elements of N, a set with an infinity of members, are matched one-to-one with all of N's subsets. Each matching must meet one of two conditions:

(1) An element is paired with a subset that includes that element. Let us call all such elements blue.

(2) An element is paired with a subset that does *not* include that element. We call all such elements red.

The red elements form a subset of our initial set N. Can this subset be matched to a blue element? No, because every blue element is in its matching subset, therefore the red subset would have to include a blue element. Can the red subset be paired with a red element? No, because the red element would then be included in its subset and would therefore be blue. Since the red subset cannot be matched to either a red or blue element of N, we have constructed a subset of N that is not paired with any element of N. No set, even if infinite, can be put into one-to-one correspondence with its subsets. If n is a transfinite number, then 2^n—by definition it is the number of subsets of n—must be a higher order of infinity than n.

Georg Cantor, the founder of set theory, used the term aleph-null for the lowest transfinite number. It is the cardinal number of the set of all integers, and for that reason is often called a "countable infinity." Any set that can be matched one-to-one with the counting numbers, such as the set of integral fractions, is said to be a countable or aleph-null set. Cantor showed that when 2 is raised to the power of aleph-null—giving the number of subsets of the integers—the result is equal to the cardinal number of the set of all real numbers (rational or irrational), called the "power of the continuum," or c. It is the cardinal number of all points on a line. The line may be a segment of any finite length, a ray with a beginning but no end, or a line going to infinity in both directions. Figure 18 shows three intuitively obvious geometrical proofs that all three kinds of line have the same number of points. The slant lines projected from point P indicate how all points on the line segment AB can be put into one-to-one correspondence with all points on the longer segment, on a ray, and on an endless line.

The red-blue proof outlined above (Cantor published it in 1890) of course generates an infinite hierarchy of transfinite

Figure 18

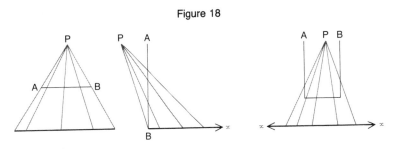

Number of points on a line segment AB is the same as on a
longer line segment *(left)*, a ray *(center)* and a line *(right)*

numbers. The ladder starts with the set of counting numbers,
aleph-null, next comes c, then all the subsets of c, then all the
subsets of all the subsets of c, and so on. The ladder can also
be expressed like this:

aleph-null, c, 2^c, 2^{2^c}, $2^{2^{2^c}}$, ...

Cantor called c "aleph-one" because he believed that no
transfinite number existed between aleph-null and c. And he
called the next number aleph-two, the next aleph-three, and so
on. For many years he tried unsuccessfully to prove that c was
the next higher transfinite number after aleph-null, a conjecture that came to be called the "continuum hypothesis." We
now know, thanks to proofs by Kurt Gödel and Paul Cohen,
that the conjecture is undecidable within standard set theory,
even when strengthened by the axiom of choice. We can assume without contradiction that Cantor's alephs catch all transfinite numbers, or we can assume, also without contradiction,
a non-Cantorian set theory in which there is an infinity of
transfinite numbers between any two adjacent entries in Cantor's ladder. (See Chapter 3 of my *Mathematical Carnival* for a
brief, informal account of this.)

Cantor also tried to prove that the number of points on a
square is the next higher transfinite cardinal after c. In 1877
he astounded himself by finding an ingenious way to match all
the points of a square to all the points of a line segment. Imagine a square one mile on a side, and a line segment one inch
long [*see Figure* 19]. On the line segment every point from 0 to
1 is labeled with an infinite decimal fraction: The point corresponding to the fractional part of pi is .14159 . . . , the point
corresponding to 1/3 is .33333 . . . and so on. Every point is
represented by a unique string of aleph-null digits, and every
possible aleph-null string of digits represents a unique point on
the line segment. (A slight difficulty arises from the fact that a

Figure 19

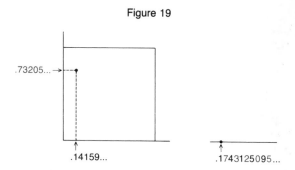

Points in square mile and on line segment

fraction such as .5000 . . . is the same as .4999 . . . , but it is easily overcome by dodges we need not go into here.)

Now consider the square mile. Using a Cartesian coordinate system, every point on the square has unique x and y coordinates, each of which can be represented by an endless decimal fraction. The illustration shows a point whose x coordinate is the fractional part of pi and whose y coordinate is the fractional part of the square root of 3, or .73205. . . . Starting with the x coordinate, alternate the digits of the two numbers: .1743125095. . . . The result is an endless decimal labeling a unique point on the line segment. Clearly this can be done with every point on the square. It is equally obvious that the mapping procedure can be reversed: we can select any point on the line segment and, by taking alternate digits of its infinite decimal, can split it into two endless decimals that as coordinates label a unique point on the square. (Here we must recognize and overcome the subtle fact that, for example, the following three distinct points on the segment—.449999 . . . , .459090 . . . , and .540909 . . .—all map the same point [½, ½] in the square.) In this way the points of any square can be put into one-to-one correspondence with the points on any line segment; therefore the two sets are equivalent and each has the cardinal number c.

The proof extends easily to a cube (by interlacing three coordinates), or to a hypercube of n dimensions (by interlacing n coordinates). Other proofs show that c also numbers the points in an infinite space of any finite number of dimensions, even an infinite space of aleph-null dimensions.

Cantor hoped that his transfinite numbers would distinguish the different orders of space but, as we have seen, he himself proved that this was not the case. Mathematicians later showed that it is the topological way the points of space go together that distinguishes one space from another. The matchings in the previous paragraphs are not continuous; that is, points close together on, for instance, the line are not necessarily close together on the square, and vice versa. Put another way, you cannot continuously deform a line to make it a square, or a square to make it a cube, or a cube to make a hypercube, and so on.

Is there a set in mathematics that corresponds to 2^c? Of course we know it is the number of all subsets of the real numbers, but does it apply to any familiar set in mathematics? Yes, it is the set of all real functions of x, even the set of all real one-valued functions. This is the same as the number of all possible permutations of the points on a line. Geometrically it is all the curves (including discontinuous ones) that can be drawn on a plane or even a small finite portion of a plane the size, say, of a postage stamp. As for 2 to the power of 2^c, no one has yet found a set, aside from the subsets of 2^c, equal to it. Only aleph-null, c, and 2^c seem to have an application outside the higher reaches of set theory. As George Gamow once said, "we find ourselves here in a position exactly opposite to that of . . . the Hottentot who had many sons but could not count beyond three." There is an endless ladder of transfinite numbers, but most mathematicians have only three "sons" to count with them. This has not prevented philosophers from trying to find metaphysical interpretations for the transfinite numbers. Cantor himself, a deeply religious man, wrote at length on such matters. In the United States, Josiah Royce was the philosopher who made the most extensive use of Cantor's alephs, particularly in his work *The World and the Individual*.

The fact that there is no highest or final integer is involved in a variety of bewildering new paradoxes. Known as supertasks, they have been much debated by philosophers of science since they were first suggested by the mathematician Hermann Weyl. For instance, imagine a lamp (called the Thomson lamp after James F. Thomson, who first wrote about it) that is turned off and on by a push-button switch. Starting at zero time, the lamp is on for 1/2 minute, then it is off for 1/4 minute, then on for 1/8 minute and so on. Since the sum of this halving series, $1/2 + 1/4 + 1/8 + \ldots$, is 1, at the end of one minute the switch will have been moved aleph-null times. Will the lamp be on or off?

Everyone agrees that a Thomson lamp cannot be constructed. Is such a lamp logically conceivable or is it nonsense to discuss it in the abstract? One of Zeno's celebrated paradoxes concerns a constant-speed runner who goes half of a certain distance in 1/2 minute, a fourth of the distance in the next

1/4 minute, an eighth of the distance in the next 1/8 minute and so on. At the end of one minute he has had no difficulty reaching the last point of the distance. Why, then, cannot we say that at the end of one minute the switch of the Thomson lamp has made its last move? The answer is that the lamp must then be on or off and this is the same as saying that there is a last integer that is either even or odd. Since the integers have no last digit, the lamp's operation seems logically absurd.

Another supertask concerns an "infinity machine" that calculates and prints the value of pi. Each digit is printed in half the time it takes to print the preceding one. Moreover, the digits are printed on an idealized tape of finite length, each digit having half the width of the one before it. Both the time and the width series converge to the same limit, so that in theory one might expect the pi machine, in a finite time, to print all the digits of pi on a piece of tape. But pi has no final digit to print, and so again the supertask seems self-contradictory.

One final example: Max Black of Cornell University imagines a machine that transfers a marble from tray A to tray B in one minute and then rests for a minute as a second machine returns the marble to A. In the next half-minute the first machine moves the marble back to B; then it rests for a half-minute as the other machine returns it to A. This continues, in a halving time series, until the machines' movements become, as Black puts it, a "grey blur." At the end of four minutes each machine has made aleph-null transfers. Where is the marble? Once more, the fact that there is no last integer to be odd or even seems to rule out the possibility, even in principle, of such a supertask. (The basic articles on supertasks, by Thomson, Black and others, are reprinted in Wesley C. Salmon's 1970 paperback anthology *Zeno's Paradoxes*.)

One is tempted to say that the basic difference between supertasks and Zeno's runner is that the runner moves continuously whereas the supertasks are performed in discrete steps that form an aleph-null set. The situation is more complicated than that. Adolph Grünbaum, in *Modern Science and Zeno's Paradoxes*, argues convincingly that Zeno's runner could also complete his run by what Grünbaum calls a "staccato" motion of aleph-null steps. The staccato runner goes the first half of his distance in 1/4 minute, rests 1/4 minute, goes half of the remaining distance in 1/8 minute, rests 1/8 minute and so on. When he is running, he moves twice as fast as his "legato" counterpart, but his overall average speed is the same, and it is always less than the velocity of light. Since the pauses of the staccato runner converge to zero, at the end of one minute he too will have reached his final point just as an ideal bouncing ball comes to rest after an infinity of discrete bounces. Grünbaum finds no logical objection to the staccato run, even though it cannot be carried out in practice. His attitudes toward the supertasks are complex and controversial. He regards infinity machines of certain designs as being logically impossible and yet in most cases, with suitable qualifications, he defends them as logically consistent variants of the staccato run.

These questions are related to an old argument to the effect that Cantor was mistaken in his claim that aleph-null and c are different orders of infinity. The proof is displayed in Figure 20. The left side is an endless list of integers in serial order.

Figure 20

INTEGERS	DECIMAL FRACTIONS
1	.1
2	.2
3	.3
.	.
.	.
.	.
10	.01
11	.11
12	.21
.	.
.	.
.	.
100	.001
101	.101
.	.
.	.
.	.
1234	.4321
.	.
.	.
.	.

Fallacious proof concerning two alephs

Each is matched with a number on the right that is formed by reversing the order of the digits and putting a decimal in front of them. Since the list on the left can go to infinity, it should eventually include every possible sequence of digits. If it does, the numbers on the right will also catch every possible sequence and therefore will represent all real numbers between 0 and 1. The real numbers form a set of size c. Since this set can be put in one-to-one correspondence with the integers, an aleph-null set, the two sets appear to be equivalent.

I would be ashamed to give this proof were it not for the fact that every year or so I receive it from a correspondent who has rediscovered it and convinced himself that he has demolished Cantorian set theory. Readers should have little difficulty seeing what is wrong.

ANSWERS

The fundamental error in the false proof that the counting numbers can be matched one-to-one with the real numbers is that, no matter how long the list of integers on the left (and their mirror reversals on the right), no number with aleph-null digits will ever appear on each side. As a consequence no irrational decimal fraction will be listed on the right. The mirror reversals of the counting numbers, with a decimal point in front of each, form no more than a subset of the integral fractions between 0 and 1. Not even 1/3 appears in this subset because its decimal form requires aleph-null digits. In brief, all that is proved is the well-known fact that the counting numbers can be matched one-to-one with a subset of integral fractions.

The false proof reminds me of a quatrain I once perpetrated:

> Pi vs *e*
>
> Pi goes on and on and on . . .
> And *e* is just as cursed.
> I wonder: Which is larger
> When their digits are reversed?

ADDENDUM

Among physicists, no one objected more violently to Cantorian set theory than Percy W. Bridgman. In *Reflections of a Physicist* (1955) he says he "cannot see an iota of appeal" in Cantor's proof that the real numbers form a set of higher infinity than the integers. Nor can he find paradox in any of Zeno's arguments because he is unable to think of a line as a set of points (see the *Clerihew* by Lindon that I used as an epigraph) or a time interval as a set of instants.

"A point is a curious thing," he wrote in *The Way Things Are* (1959), "and I do not believe that its nature is appreciated, even by many mathematicians. A line is not composed of points in any real sense. . . . We do not construct the line out of points, but, given the line, we may construct points on it. 'All the points on the line' has the same sort of meaning that the 'entire line' has. . . . We *create* the points on a line just as we create the numbers, and we identify the points by the numerical values of the coordinates."

Merwin J. Lyng, in *The Mathematics Teacher* (April 1968, page 393), gives an amusing variation of Black's moving-marble supertask. A box has a hole at each end: Inside the box a rabbit sticks his head out of hole *A*, then a minute later out of hole *B*, then a half-minute later out of hole *A*, and so on. His students concluded that after two minutes the head is sticking out of both holes, "but practically the problem is not possible unless we split hares."

For what it is worth, I agree with those who believe that paradoxes such as the staccato run can be stated without contradiction in the language of set theory, but as soon as any element is added to the task that involves a highest integer, you add something not permitted, therefore you add only nonsense. There is nothing wrong in the abstract about an ideal bouncing ball coming to rest, or a staccato moving point reaching a goal, but nothing meaningful is added if you assume that at each bounce the ball changes color, alternating red and blue; then ask what color it is when it stops bouncing, or if the staccato runner opens and shuts his mouth at each step and you ask if it is open or closed at the finish.

A number of readers called my attention to errors in this chapter, as I first wrote it as a column, but I wish particularly to thank Leonard Gillman, of the University of Texas at Austin for reviewing the column and suggesting numerous revisions that have greatly simplified and improved the text.

BIBLIOGRAPHY

Infinity, an Essay in Metaphysics. José Benardete. Clarendon Press, 1964.

"The Achilles Paradox and Transfinite Numbers." C. David Gruender. *British Journal for the Philosophy of Science*, Vol. 17, November 1966, pages 219–231.

Modern Science and Zeno's Paradoxes. Adolf Grünbaum. Wesleyan University Press, 1967.

"Are 'Infinity Machines' Paradoxical?" Adolf Grünbaum. *Science*, Vol. 159, January 26, 1968, pages 396–406.

"Can an Infinitude of Operations Be Performed in a Finite Time?" Adolf Grünbaum. *British Journal for the Philosophy of Science*, Vol. 20, October 1969, pages 203–218.

Zeno's Paradoxes. Wesley C. Salmon, ed. Bobbs-Merrill, 1970.

"Zeno, Aristotle, Weyl and Shuard: Two-and-a-half millenia of worries over number." C. W. Kilmister. *The Mathematical Gazette*, Vol. 64, October 1980, pages 149–158.

5

NONTRANSITIVE DICE

AND OTHER PROBABILITY PARADOXES

Probability theory abounds in paradoxes that wrench common sense and trap the unwary. In this chapter we consider a startling new paradox involving the relation called transitivity and a group of paradoxes stemming from the careless application of what is called the principle of indifference.

Transitivity is a binary relation such that if it holds between A and B and between B and C, it must also hold between A and C. A common example is the relation "heavier than." If A is heavier than B and B is heavier than C, then A is heavier than C. The three sets of four dice shown "unfolded" in Figure 21 were designed by Bradley Efron, a statistician at Stanford University, to dramatize some recent discoveries about a general class of probability paradoxes that violate transitivity. With any of these sets of dice you can operate a betting game so contrary to intuition that experienced gamblers will find it almost impossible to comprehend even after they have completely analyzed it.

The four dice at the top of the illustration are numbered in the simplest way that provides the winner with the maximum advantage. Allow someone to pick any die from this set. You then select a die from the remaining three. Both dice are tossed and the person who gets the highest number wins. Surely, it seems, if your opponent is allowed the first choice of a die before each contest, the game must either be fair or favor your opponent. If at least two dice have equal and maximum probabilities of winning, the game is fair because if he picks one such die, you can pick the other; if one die is better than the other three, your opponent can always choose that die and win more than half of the contests. This reasoning is com-

Figure 21

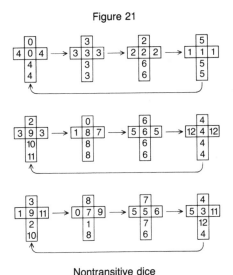

Nontransitive dice

pletely wrong. The incredible truth is that regardless of which die he picks you can always pick a die that has a 2/3 probability of winning, or two-to-one odds in your favor!

The paradox (insofar as it violates common sense) arises from the mistaken assumption that the relation "more likely to win" must be transitive between pairs of dice. This is not the case with any of the three sets of dice. In each set the relation "more likely to win" is indicated by an arrow that points to the losing die. Die A beats B, B beats C, C beats D—and D beats A! In the first set the probability of winning with the indicated die of each pair is 2/3. This is easily verified by listing the 36 possible throws of each pair, then checking the 24 cases in which one die bears the highest number.

The other two sets of four dice, also designed by Efron, have the same nontransitive property but fewer numbers are repeated in order to make an analysis of the dice more difficult. In the second set the probability of winning with the indicated die is also 2/3. Because ties are possible with the third set it must be agreed that ties will be broken by rolling again. With this procedure the winning probability for each of the four pairings in the third set is 11/17, or .647.

It has been proved, Efron writes, that 2/3 is the greatest possible advantage that can be achieved with four dice. For three sets of numbers the maximum advantage is .618, but this cannot be obtained with dice because the sets must have more than six numbers. If more than four sets are used (numbers to be randomly selected within each set), the possible advantage approaches a limit of 3/4 as the number of sets increases.

A fundamental principle in calculating probabilities such as dice throws is one that goes back to the beginnings of classical probability theory in the 18th century. It was formerly called "the principle of insufficient reason" but is now known as "the principle of indifference," a crisper phrase coined by John Maynard Keynes in *A Treatise on Probability*. (Keynes is best known as an economist, but his book on probability has become a classic. It had a major influence on the inductive logic of Rudolf Carnap.) The principle is usually stated as follows: If you have no grounds whatever for believing that any one of n mutually exclusive events is more likely to occur than any other, a probability of $1/n$ is assigned to each.

For example, you examine a die carefully and find nothing that favors one side over another, such as concealed loads, noncubical shape, beveling of certain edges, stickiness of certain sides and so on. You assume that there are six equally probable ways the cube can fall; therefore you assign a probability of 1/6 to each. If you toss a penny, or play the Mexican game of betting on which of two sugar cubes a fly will alight on first, your ignorance of any possible bias prompts you to assign a probability of 1/2 to each of the two outcomes. In none of these samples do you feel obligated to make statistical, empirical tests. The probabilities are assigned a priori. They are based on symmetrical features in the structures and forces involved. The die is a regular solid, the probability of the penny's balancing on its edge is virtually zero, there is no reason for a fly to prefer one sugar cube to another and so on. Ultimately, of course, your analysis rests on empirical grounds, since only experience tells you, say, that a weighted die face would affect the odds, whereas a face colored red (with the others blue) would not.

Some form of the principle of indifference is indispensable in probability theory, but it must be carefully qualified and applied with extreme caution to avoid pitfalls. In many cases the traps spring from a difficulty in deciding on what are the equally probable cases. Suppose, for instance, you shuffle a packet of four cards—two red, two black—and deal them face down in a row. Two cards are picked at random, say by placing a penny on each. What is the probability that those two cards are the same color?

One person reasons: "There are three equally probable cases. Either both cards are black, both are red or they are different colors. In two cases the cards match, therefore the matching probability is 2/3."

"No," another person counters, "there are *four* equally probable cases. Either both cards are black, both are red, card *x* is black and *y* is red or *x* is red and *y* is black. More simply, the cards either match or they do not. In each way of putting it the matching probability clearly is 1/2."

The fact is that both people are wrong. (The correct probability will be given in the Answer Section. Can the reader calculate it?) Here the errors arise from a failure to identify correctly the equally probable cases. There are, however, more confusing paradoxes—actually fallacies—in which the principle of indifference seems intuitively to be applicable, whereas it actually leads straight to a logical contradiction. Cases such as these result when there are no positive reasons for believing *n* events to be equally probable and the assumption of equiprobability is therefore based entirely, or almost entirely, on ignorance.

For example, someone tells you: "There is a cube in the next room whose size has been selected by a randomizing device. The cube's edge is not less than one foot or more than three feet." How would you estimate the probability that the cube's edge is between one and two feet as compared with the probability that it is between two and three feet? In your total ignorance of additional information, is it not reasonable to invoke the principle of indifference and regard each probability as 1/2?

It is not. If the cube's edge ranges between one and two feet, its volume ranges between 1^3, or one, cubic foot and 2^3, or eight, cubic feet. But in the range of edges from two to three feet, the volume ranges between 2^3 (eight) and 3^3 (27) cubic feet—a range almost three times the other range. If the principle of indifference applies to the two ranges of edges, it is violated by the equivalent ranges of volume. You were not told how the cube's "size" was randomized, and since "size" is ambiguous (it could mean either the cube's edge or its volume) you have no clues to guide your guessing. If the cube's edge was picked at random, the principle of indifference does indeed apply. It is also applicable if you are told that the cube's volume was picked at random, but of course you then have to assign a probability of 1/2 to each of the two ranges from one to 14 and from 14 to 27 cubic feet, and to the corresponding ranges for the cube's edge. If the principle applies to the edge, it cannot apply to the volume without contradiction, and vice versa. Since you do not know how the size was selected, any application of the principle is meaningless.

Carnap, in attacking an uncritical use of the principle in Harold Jeffreys' *Theory of Probability,* gives the following example of its misuse. You know that every ball in an urn is blue, red or yellow, but you know nothing about how many balls of each color are in the urn. What is the probability that the first ball taken from the urn will be blue? Applying the principle of indifference, you say it is 1/2. The probability that it is not blue must also be 1/2. If it is not blue, it must be red or yellow, and because you know nothing about the number of red or yellow balls, those colors are equally probable. Therefore you assign to red a probability of 1/4. On the other hand, if you begin by asking for the probability that the first ball will be red, you must give red a probability of 1/2 and blue a probability of 1/4, which contradicts your previous estimates.

It is easy to prove along similar lines that there is life on Mars. What is the probability that there is simple plant life on Mars? Since arguments on both sides are about equally cogent, we answer 1/2. What is the probability that there is simple animal life on Mars? Again, 1/2. Now we are forced to assert that the probability of there being "either plant or animal life" on Mars is $1/2 + 1/2 = 1$, or certainty, which is absurd. The philosopher Charles Sanders Peirce gave a similar argument that seems to show that the hair of inhabitants on Saturn had to be either of two different colors. Many variants of this fallacy can be found in Chapter 4 of Keynes's book. It is easy to invent others.

In the history of metaphysics the most notorious misuse of the principle surely was by Blaise Pascal, who did pioneer work on probability theory, in a famous argument that became known as "Pascal's wager." A few passages from the original and somewhat lengthy argument (in Pascal's *Pensées,* Thought 233) are worth quoting:

> "God is, or he is not." To which side shall we incline? Reason can determine nothing about it. There is an infinite gulf fixed between us. A game is playing at the extremity of this infinite distance in which heads or tails may turn up. What will you wager? There is no reason for backing either one or the other, you cannot reasonably argue in favor of either. . . .
>
> Yes, but you must wager. . . . Which will you choose? . . . Let us weigh the gain and the loss in choosing "heads" that God is. . . . If you gain, you gain all. If you lose, you lose nothing. Wager, then, unhesitatingly that he is.

Lord Byron, in a letter, rephrased Pascal's argument effectively: "Indisputably, the firm believers in the Gospel have a great advantage over all others, for this simple reason—that, if true, they will have their reward hereafter; and if there be no hereafter, they can be but with the infidel in his eternal sleep, having had the assistance of an exalted hope through life, without subsequent disappointment, since (at the worst for them) out of nothing nothing can arise, not even sorrow." Similar passages can be found in many contemporary books of religious apologetics.

Pascal was not the first to insist in this fashion that faith in Christian orthodoxy was the best bet. The argument was clearly stated by the fourth-century African priest Arnobius the Elder, and non-Christian forms of it go back to Plato. This is not the place, however, to go into the curious history of defenses and criticisms of the wager. I content myself with mentioning Denis Diderot's observation that the wager applies with equal force to other major faiths such as Islam. The mathematically interesting aspect of all of this is that Pascal likens the outcome of his bet to the toss of a coin. In other words, he explicitly invokes the principle of indifference to a situation in which its application is mathematically senseless.

The most subtle modern reformulation of Pascal's wager is by William James, in his famous essay *The Will to Believe,* in which he argues that philosophical theism is a better gamble than atheism. In a still more watered-down form it is even used occasionally by humanists to defend optimism against pessimism at a time when the extinction of the human race seems as likely in the near future as its survival.

"While there is a chance of the world getting through its troubles," says the narrator of H. G. Wells's little read novel *Apropos of Dolores,* "I hold that a reasonable man has to behave as though he was sure of it. If at the end your cheerfulness is not justified, at any rate you will have been cheerful."

ANSWERS

The probability that two randomly selected cards, from a set of two red and two black cards, are the same color is 1/3. If you list the 24 equally probable permutations of the four cards, then pick any two positions (for example second and fourth cards), you will find eight cases in which the two cards match in color. One way to see that this probability of 8/24 or 1/3 is correct is to consider one of the two chosen cards. Assume that it is red. Of the remaining three cards only one is red, and so the probability that the second chosen card will be red is 1/3. Of course, the same argument applies if the first card is black. Most people guess that the odds are even, when actually they are two to one in favor of the cards' having different colors.

ADDENDUM

The following letter, from S. D. Turner, contains some surprising information:

> Your bit about the two black and two red cards reminds me of an exercise I did years ago, which might be called *N*-Card Monte. A few cards, half red, half black, or nearly so, are shown face up by the pitchman, then shuffled and dealt face down. The sucker is induced to bet he can pick two of the same color.
>
> The odds will always be against him. But because the sucker will make erroneous calculations (like the 2/3 and 1/2 in your 2:2 example), or for other reasons, he will bet. The pitchman can make a plausible spiel to aid this: "Now, folks, you don't need to pick two blacks, and you don't need to pick two reds. If you draw either pair you win!"
>
> The probability of getting two of the same color, where there are R reds and B blacks, is:
>
> $$(1) \quad P = \frac{R^2 + B^2 - (R + B)}{(R + B)(R + B - 1)}$$
>
> This yields the figures in the table [*see Figure* 22], one in lowest-terms fractions, the other in decimal. Only below and to the left of the stairstep line does the sucker get an even break or better. But no pitchman would bother with odds more favorable to the sucker than the 1/3 probability for 2:2, or possibly the 2/5 for 3:3.
>
> Surprisingly, the two top diagonal lines are identical. That is, if you are using equal reds and blacks, odds are not changed if a card is removed before the two are selected! In your example of 2:2, the probability is 1/3 and it is also 1/3 when starting with 2:1 (as is evident because the one card not selected can be any one of the three). The generality of this can be shown thus: If $B = R$ and $B = R-1$ are substituted into (1), the result in each case is $R-1/2R-1$.

Some readers sent detailed explanations of why the arguments behind the fallacies that I described were wrong, apparently not realizing that these fallacies were intended to be howlers based on the misuse of the principle of indifference.

Figure 22

		Red Cards									
Black Cards		1	2	3	4	5	6	7	8	... 13	26
	2	$\frac{1}{3}$	$\frac{1}{3}$								
	3	$\frac{1}{2}$	$\frac{2}{5}$	$\frac{2}{5}$							
	4	$\frac{3}{5}$	$\frac{7}{15}$	$\frac{3}{7}$	$\frac{3}{7}$						
	5	$\frac{2}{3}$	$\frac{11}{21}$	$\frac{13}{28}$	$\frac{4}{9}$	$\frac{4}{9}$					
	6	$\frac{5}{7}$	$\frac{4}{7}$	$\frac{1}{2}$	$\frac{7}{15}$	$\frac{5}{11}$	$\frac{5}{11}$				
	7	$\frac{3}{4}$	$\frac{11}{18}$	$\frac{8}{45}$	$\frac{27}{35}$	$\frac{31}{66}$	$\frac{6}{13}$	$\frac{6}{13}$			
	8	$\frac{7}{9}$	$\frac{29}{45}$	$\frac{31}{55}$	$\frac{17}{33}$	$\frac{19}{39}$	$\frac{43}{91}$	$\frac{7}{15}$	$\frac{7}{15}$		
	⋮										
	13									$\frac{12}{25}$	
	26										$\frac{25}{51}$
	2	.333	.333								
	3	.500	.400	.400							
	4	.600	.466	.429	.429						
	5	.667	.524	.465	.444	.444					
	6	.714	.572	.500	.467	.455	.455				
	7	.750	.611	.533	.491	.470	.462	.462			
	8	.778	.645	.564	.515	.488	.472	.466	.466		
	⋮										
	13									.480	
	26										.490

Probability of drawing two cards of the same color

Several readers correctly pointed out that although Pascal did invoke the principle of indifference by referring to a coin flip in his famous wager, the principle is not essential to his argument. Pascal posits an infinite gain for winning a bet in which the loss (granting his assumptions) would always be finite regardless of the odds.

Efron's nontransitive dice aroused almost as much interest among magicians as among mathematicians. It was quickly perceived that the basic idea generalized to k sets of n-sided dice, such as dice in the shapes of regular octahedrons, dodecahedrons, icosahedrons, or cylinders with n flat sides. The game also can be modeled by k sets of n-sided tops, spinners with n numbers on each dial, and packets of n playing cards.

Karl Fulves, in his magic magazine *The Pallbearers Review* (January 1971) proposed using playing cards to model Efron's dice. He suggested the following four packets: 2, 3, 4, 10, J, Q; 1, 2, 8, 9, 9, 10; 6, 6, 7, 7, 8, 8; and 4, 5, 5, 6, Q, K. Suits are irrelevant. First player selects a packet, shuffles it, and draws a card. Second player does the same with another packet. If the chosen cards have the same value, they are replaced and two more cards drawn. Ace is low, and high card wins. This is based on Efron's third set of dice where the winning probability, if the second player chooses properly, is 11/17. To avoid giving away the cyclic sequence of packets, each could be placed in a container (box, cup, tray, etc.) with the containers secretly marked. Before each play, the containers would be randomly mixed by the first player while the second player turned his back. Containers with numbered balls or counters could of course be substituted for cards.

In the same issue of *The Pallbearers Review* cited above, Columbia University physicist Shirley Quimby proposed a set of four dice with the following faces:

$$3, 4, 5, 20, 21, 22$$
$$1, 2, 16, 17, 18, 19$$
$$10, 11, 12, 13, 14, 15$$
$$6, 7, 8, 9, 23, 24$$

Note that numbers 1 through 24 are used just once each in this elegant arrangement. The dice give the second player a winning probability of 2/3. If modeled with 24 numbered cards, the first player would select one of the four packets, shuffle, then draw a card. The second player would do likewise, and high card wins.

R. C. H. Cheng, writing from Bath University, England, proposed a novel variation using a single die. On each face are numbers 1 through 6, each numeral a different color. Assume that the colors are the rainbow colors red, orange, yellow, green, blue, and purple. The chart below shows how the numerals are colored on each face.

Face	Red	Orange	Yellow	Green	Blue	Purple
A	1	2	3	4	5	6
B	6	1	2	3	4	5
C	5	6	1	2	3	4
D	4	5	6	1	2	3
E	3	4	5	6	1	2
F	2	3	4	5	6	1

The game is played as follows: The first player selects a color, then the second player selects another color. The die is rolled and the person whose color has the highest value wins. It is easy to see from the chart that if the second player picks the adjacent color on the right—the sequence is cyclic, with red to the "right" of purple—the second player wins five out of six times. In other words, the odds are 5 to 1 in his favor!

To avoid giving away the sequence of colors, the second player should occasionally choose the second color to the right, where his winning odds are 4 to 2, or the color third to the right where the odds are even. Perhaps he should even, on rare occasions, take the fourth or fifth color to the right where odds against him are 4 to 2 and 5 to 1 respectively. Mel Stover has suggested putting the numbers and colors on a 6-sided log instead of a cube.

This, too, models nicely with 36 cards, formed in six piles, each bearing a colored numeral. The chart's pattern is obvious, and easily applied to n^2 cards, each with numbers 1 through n, and using n different colors. In presenting it as a betting game you should freely display the faces of each packet to show that all six numbers and all six colors are represented. Each packet is shuffled and placed face down. The first player is "generously" allowed first choice of a color, and to select any packet. The color with the highest value in that packet is the winner. In the general case, as Cheng pointed out in his 1971 letter, the second player can always choose a pile that gives him a probability of winning equal to $(n-1)/n$.

A simpler version of this game uses 16 playing cards. The four packets are:

AS, JH, QC, KD
KS, AH, JC, QD
QS, KH, AC, JD
JS, QH, KC, AD

Ace here is high, and the cyclic sequence of suits is spades, hearts, clubs, diamonds. The second player wins with 3 to 1 odds by choosing the next adjacent suit, and even odds if he goes to the next suit but one.

These betting games are all variants of nontransitive voting paradoxes, about which there is extensive literature.

BIBLIOGRAPHY

A Treatise on Probability. John Maynard Keynes. Macmillan, 1921. Harper & Row paperback reprint, 1962.

"Statistical and Inductive Probability." Rudolf Carnap in *The Structure of Scientific Thought*, by Edward H. Madden, ed. Houghton Mifflin, 1960.

On Pascal's Wager:

"Pascal's Wager." James Cargile. *Philosophy: The Journal of the Royal Institute of Philosophy*, Vol. 41, July 1966, pages 250–257.

"Deciding for God—the Bayesian Support of Pascal's Wager." Merle B. Turner in *Philosophy and Phenomenological Research*, Vol. 29, September 1968, pages 84–90.

The Emergence of Probability: A Philosophical Study of Early Ideas About Probability, Induction, and Statistical Inference. Ian Hacking. Cambridge University Press, 1976.

On Nontransitive Betting Games:

"A Dice Paradox." Gene Finnell. *Epilogue,* July 1971, pages 2–3. This is another magic magazine published by Karl Fulves.

"Nontransitive Dominance." Richard Tenney and Caxton Foster. *Mathematics Magazine,* Vol. 49. May–June 1976, pages 115–120.

"Mathematical Games." Martin Gardner. *Scientific American,* October 1974.

6

GEOMETRICAL FALLACIES

"Holmes," I cried, "this is impossible."
"Admirable!" he said. "A most illuminating
remark. It *is* impossible as I state it, and
therefore I must in some respect have stated
it wrong. . . ."

—SIR ARTHUR CONAN DOYLE,
The Adventure of the Priory School

It is commonly supposed that Euclid, the ancient Greek geometer, wrote only one book, his classic *Elements of Geometry*. Actually he wrote at least a dozen, including treatises on music and branches of physics, but only five of his works survived. One of his lost books was a collection of geometric fallacies called *Pseudaria*. Alas, there are no records of what it contained. It probably discussed illicit proofs that led to absurd theorems but in which the errors were not immediately apparent.

Since Euclid's time hundreds of amusing examples of geometric fallacies have been published, some of them genuine mistakes and some deliberately contrived. This month we consider five of the best. All are theorems that could have been in Euclid's *Pseudaria*, since none requires more than a knowledge of elementary plane geometry to follow their steps down the garden path to the false conclusion. (Q.E.D.: Quite Entertainingly Deceptive.) The reader is urged to examine each proof carefully, step by step, to see if he can discern exactly where the proof goes wrong before the errors are revealed in the Answer Section.

Figure 23

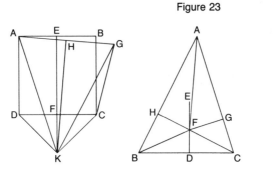
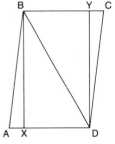

Obtuse angle equals right angle
All triangles are isoceles
ABCD is a parallelogram

THEOREM 1: AN OBTUSE ANGLE IS SOMETIMES EQUAL TO A RIGHT ANGLE. This was one of Lewis Carroll's favorites. Figure 23, *left*, reproduces Carroll's diagram and labeling. I know of no better way for a high school geometry teacher to convey the importance of deductive rigor than to chalk this diagram on the blackboard and challenge a class to find where the fallacy lies. The construction and proof are described by Carroll as follows (I quote from *The Lewis Carroll Picture Book*, edited by Stuart Dodgson Collingwood, London, 1899; reprinted in the Dover paperback *Diversions and Digressions of Lewis Carroll*, 1961):

Let *ABCD* be a square. Bisect *AB* at *E*, and through *E* draw *EF* at right angles to *AB*, and cutting *DC* at *F*. Then *DF* = *FC*.

From *C* draw *CG* = *CB*. Join *AG*, and bisect it at *H*, and from *H* draw *HK* at right angles to *AG*.

Since *AB*, *AG* are not parallel, *EF*, *HK* are not parallel. Therefore they will meet, if produced. Produce *EF*, and let them meet at *K*. Join *KD*, *KA*, *KG*, and *KC*.

The triangles *KAH*, *KGH* are equal, because *AH* = *HG*, *HK* is common, and the angles at *H* are right. Therefore *KA* = *KG*.

The triangles *KDF*, *KCF* are equal, because *DF* = *FC*, *FK* is common, and the angles at *F* are right. Therefore *KD* = *KC*, and angle *KDC* = angle *KCD*.

Also *DA* = *CB* = *CG*.

Hence the triangles *KDA*, *KCG* have all their sides equal. Therefore the angles *KDA*, *KCG* are equal. From these equals take the equal angles *KDC*, *KCD*. Therefore the remainders are equal: *i.e.*, the angle *GCD* = the angle *ADC*. But *GCD* is an obtuse angle, and *ADC* is a right angle.

Therefore an obtuse angle sometimes = a right angle.

Q.E.D.

THEOREM 2: EVERY TRIANGLE IS ISOSCELES. This marvelous absurdity is also in *The Lewis Carroll Picture Book*. Carroll probably came on both proofs in the first (1892) edition of W. W. Rouse Ball's *Mathematical Recreations and Essays*, where they appeared for the first time. Carroll has explained it so well that again I give his diagram [*see Figure* 23, *middle*] and quote his wording:

Let *ABC* be any triangle. Bisect *BC* at *D*, and from *D* draw *DE* at right angles to *BC*. Bisect the angle *BAC*.

(1) If the bisector does not meet *DE*, they are parallel. Therefore the bisector is at right angles to *BC*. Therefore *AB* = *AC*, i.e., *ABC* is isosceles.

(2) If the bisector meets *DE*, let them meet at *F*. Join *FB*, *FC*, and from *F* draw *FG*, *FH*, at right angels to *AC*, *AB*.

Then the triangles *AFC*, *AFH* are equal, because they have the side *AF* common, and the angles *FAG*, *AGF* equal to the angles *FAH*, *AHF*. Therefore *AH* = *AG*, and *FH* = *FG*.

Again, the triangles *BDF*, *CDF* are equal, because *BD* = *DC*, *DF* is common, and the angles at *D* are equal. Therefore *FB* = *FC*.

Again, the triangles *FHB*, *FGC* are right-angled. Therefore the square on *FB* = the squares on *FH*, *HB;* and the square on *FC* = the squares on *FG*, *GC*. But *FB* = *FC*, and *FH* = *FG*. Therefore the square on *HB* = the square on *GC*. Therefore *HB* = *GC*. Also, *AH* has been proved = to *AG*. Therefore *AB* = *AC;* i.e., *ABC* is isosceles.

Therefore the triangle *ABC* is always isosceles.

Q.E.D.

THEOREM 3: IF A QUADRILATERAL *ABCD* HAS ANGLE *A* EQUAL TO ANGLE *C*, AND *AB* EQUALS *CD*, THE QUADRILATERAL IS A PARALLELOGRAM. P. Halsey of London contributed this subtle fallacy to *The Mathematical Gazette*, October, 1959, pages 204–205. On the quadrilateral shown in Figure 23 *right,* draw *BX* perpendicular to *AD*, and *DY* perpendicular to *BC*. Join *BD*. Triangles *ABX* and *CYD* are congruent, therefore *BX* equals *DY* and *AX* equals *CY*. It follows that triangles *BXD* and *DYB* are congruent, consequently *XD* equals *YB*. Since *AB* equals *CD* and *AD* equals *BC*, the quadrilateral *ABCD* must be a parallelogram. The proof is strongly convincing, yet the theorem is false. Can the reader provide a counterexample?

THEOREM 4: PI EQUALS 2. Figure 24 is based on the familiar yin-yang symbol of the Orient. Let diameter *AB* equal 2. Since a circle's circumference is its diameter times pi, the largest semicircle, from *A* to *B*, has a length of $2\pi/2 = \pi$. The two next-

Figure 24

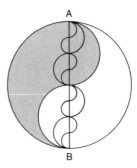

Pi equals 2

smallest semicircles, which form the wavy line that divides the yin from the yang, are each equal to $\pi/2$ and so their total length is pi. In similar fashion the sum of the four next-smallest semicircles (each $\pi/4$) also is pi, and the sum of the eight next-smallest semicircles (each $\pi/8$) also is pi. This can be continued endlessly. The semicircles grow smaller and more numerous, but they always add to pi. Clearly the wavy line approaches diameter AB as a limit. Assume that the construction is carried out an infinite number of times. The wavy line must always retain a length of pi, yet when the radii of the semicircles reach their limit of zero, they coincide with diameter AB, which has a length of 2. Consequently pi equals 2.

THEOREM 5: EUCLID'S PARALLEL POSTULATE CAN BE PROVED BY EUCLID'S OTHER AXIOMS. First, some historical background. Among Euclid's 10 axioms, his fifth postulate states that if a line A crosses two other lines, making the sum of the interior angles on the same side of A less than 180 degrees, the two lines will intersect on that side of A. A variety of seemingly unrelated theorems can be substituted for this axiom since they require it for their proof: The theorem that the interior angles of every triangle add to 180 degrees, or that a rectangle exists, or that similar noncongruent triangles exist, or that through three points not in a straight line only one circle can be drawn, and many others.

Hundreds of attempts have been made since Euclid's time to replace his cumbersome fifth postulate with one that is simpler and more intuitively obvious. The most famous became known as "Playfair's postulate" after the Scottish mathematician and physicist John Playfair. In his popular 1795 edition of Euclid's *Elements* he substituted for the fifth postulate the equivalent but more succinct statement, "Through a given point can be drawn only one line parallel to a given line." Actually this form of the fifth postulate was suggested by Proclus, in a fifth-century Greek commentary on Euclid, as well as by later mathematicians who preceded Playfair, but the parallel postulate still bears Playfair's name.

Whatever form the fifth axiom was given, it always seemed less self-evident than Euclid's other axioms, and some of the greatest mathematicians labored to eliminate it entirely by proving it on the basis of the other nine. (For a good account of this history see W. B. Frankland, *Theories of Parallelism, an Historical Critique,* Cambridge University Press, 1910.) The 18th-century French geometer Joseph Louis Lagrange was convinced that he had produced such a proof by showing (without assuming Euclid's fifth postulate) that the angles of any triangle add to a straight angle. In the middle of the first paragraph of a lecture to the French Academy on his discovery, however, he suddenly said, "Il faut que j'y songe encore" ("I shall have to think it over again"), put his papers in his pocket and abruptly left the hall.

More than a century ago it was established that it is as impossible to prove the fifth postulate as it is to trisect the angle, square the circle or duplicate the cube, yet even in this century "proofs" of the parallel axiom continue to be published. A splendid example is the heart of a 310-page book, *Euclid or Einstein,* privately printed in 1931 by Very Rev. Jeremiah Joseph Callahan, then president of Duquesne University. Since the general theory of relativity assumes the consistency of a non-Euclidean geometry, a simple way to demolish Einstein is to show that non-Euclidean geometry is contradictory. This Father Callahan proceeds to do by a lengthy, ingenious proof of the parallel postulate. It is a pleasant exercise to retrace Father Callahan's reasoning in an effort to find exactly where it goes astray. (For those who give up, the error is exposed by D. R. Ward's "A New Attempt to Prove the Parallel Postulate" in *The Mathematical Gazette,* Vol. 17, pages 101–104, May, 1933.)

A simple proof of the parallel postulate uses the diagram shown in Figure 25. *AB* is the given line and *C* the outside point. From *C* drop a perpendicular to *AB*. It can be shown, without invoking the parallel postulate, that only one such perpendicular can be drawn. Through *C* draw *EF* perpendicular to *CD*. Again, the parallel postulate is not needed to prove that this too is a unique line. Lines *EF* and *AB* are parallel. Once more, the theorem that two lines, each perpendicular to the same line, are parallel is a theorem that can be established

Figure 25

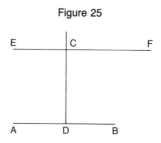

A proof of the parallel postulate

without the parallel postulate, although the proof does require other Euclidean assumptions (such as the one that straight lines are infinite in length) that do not hold in elliptic non-Euclidean geometry. Elliptic geometry does not contain parallel lines, but given Euclid's other assumptions one can assume that parallel lines do exist.

We have apparently now proved the parallel postulate. Or have we?

This and hundreds of other false proofs of Euclid's fifth axiom, or axioms equivalent to it, show how easily intuition can be deceived. It helps one to understand why it took so long for geometers to realize that the parallel postulate was independent of the others, that one may assume either that no parallel line can be drawn through the outside point, or that at least two can. (It turns out that if two can, an infinite number can.) In each case a consistent non-Euclidean geometry is constructible.

Even after non-Euclidean geometries were found to be as free of logical contradiction as Euclidean geometry, many eminent mathematicians and scientists could not believe that non-Euclidean geometry would ever have a useful application to the actual space of the universe. It is well known that Henri Poincaré argued in 1903 that if physicists ever found empirical evidence suggesting that space was non-Euclidean, it would be better to keep Euclidean geometry and change the physical laws. "Euclidean geometry, therefore," he concluded, "has nothing to fear from fresh experiments." Not so well known is the fact that Bertrand Russell and Alfred North Whitehead once voiced the same view. In 1910, in the famous 11th edition of *The Encyclopaedia Britannica,* the article on "Geometry, Non-Euclidean" is by Russell and Whitehead. If scientific observation were ever to conflict with Euclidean geometry, they assert, the simplicity of Euclidean geometry is so overwhelming that it would be preferable "to ascribe this anomaly, not to the falsity of Euclidean geometry (as applied to space), but to the falsity of the laws in question. This applies especially to astronomy."

Six years later Einstein's general theory of relativity made this statement, along with Poincaré's, hopelessly naïve. Not only does non-Euclidean geometry provide a simpler description of the space-time of general relativity; it is even possible that space may close on itself (as it does in Einstein's early model of the universe) to introduce topological properties that are in principle capable of being tested, and that could make the choice of non-Euclidean geometry as the best description of space more than a trivial matter of convention.

Russell was quick to alter the opinion expressed in the *Britannica* article but Whitehead was slow to get the point. In 1922 he wrote an embarrassing book, *The Principle of Relativity,* that attacked Einstein's use of a generalized non-Euclidean geometry (in which curvature varies from spot to spot) by arguing that simplicity demands that the geometry applied to space must be either Euclidean (Whitehead's preference) or, if the evidence warrants it, a non-Euclidean geometry in which the curvature is everywhere constant.

What is the moral of all this? Intuition is a powerful tool in mathematics and science but it cannot always be trusted. The structure of the universe, like pure mathematics itself, has a way of being much stranger than even the greatest mathematicians and physicists suspect.

ANSWERS

The errors in the fallacious geometric proofs are briefly explained as follows:

THEOREM 1. AN OBTUSE ANGLE IS SOMETIMES EQUAL TO A RIGHT ANGLE.   The mistake lies in the location of point K. When the figure is accurately drawn, K is so far below line DC that, when G and K are joined, the line falls entirely outside the original square $ABCD$. This renders the proof totally inapplicable.

THEOREM 2. EVERY TRIANGLE IS ISOSCELES.   Again the error is one of construction. F is always outside the triangle and at a point such that, when perpendiculars are drawn from F to sides AB and AC, one perpendicular will intersect one side of the triangle but the other will intersect an extension of the other side. A detailed analysis of this fallacy can be found in Eugene P. Northrop's *Riddles in Mathematics* (1944), Chapter 6.

THEOREM 3. IF A QUADRILATERAL *ABCD* HAS ANGLE *A* EQUAL TO ANGLE *C*, AND *AB* EQUALS *CD*, THE QUADRILATERAL IS A PARALLELOGRAM. The proof is correct if *X* and *Y* are each on a side of the quadrilateral or if both *X* and *Y* are on projections of the sides. It fails if one is on a side and the other is on an extension of a side, as shown in Figure 26. This figure meets the theorem's conditions but obviously is not a parallelogram.

Figure 26

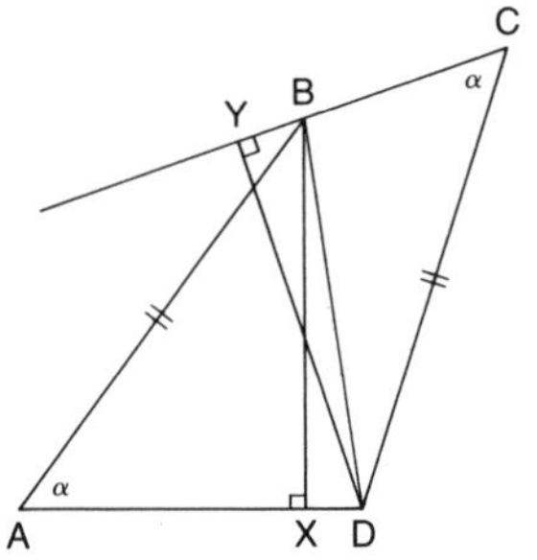

Quadrilateral-theorem counterexample

THEOREM 4. PI EQUALS 2. It is true that as the semicircles are made smaller their radii approach zero as a limit and therefore the wavy line can be made as close to the diameter of the large circle as one pleases. At no step, however, do the semicircles alter their *shape*. Since they always remain semicircles, no matter how small, their total length always remains pi. The fallacy is an excellent example of the fact that the elements of a converging infinite series may retain properties quite distinct from those of the limit itself.

THEOREM 5. EUCLID'S PARALLEL POSTULATE CAN BE PROVED BY EUCLID'S OTHER AXIOMS. The proof is valid in showing that one line can be constructed through *C* that is parallel to *AB*, but it fails to prove that there is only one such parallel. There are many other methods of constructing a parallel line through *C;* the proof does not guarantee that all these parallels are the *same* line. Indeed, in hyperbolic non-Euclidean geometry an infinity of such parallels can be drawn through *C*, a possibility that can be excluded only by adopting Euclid's fifth postulate or one equivalent to it. Elliptic non-Euclidean geometry, in which *no* parallel can be drawn through *C*, is made possible by discarding, along with the fifth postulate, certain other Euclidean assumptions.

BIBLIOGRAPHY

Fallacies in Mathematics. E. A. Maxwell, Cambridge University Press, 1959.

Whitehead's Philosophy of Science. Robert M. Palter. University of Chicago Press, 1960. Contains a thorough discussion of Whitehead's controversy with Einstein.

Lapses in Mathematical Reasoning. V. M. Bradis, V. L. Minkovskii, and A. K. Kharcheva. Pergamon, 1963.

Mistakes in Geometric Proofs. Ya. S. Dubnov. D. C. Heath, 1963.

7

THE COMBINATORICS

OF PAPER FOLDING

The easiest way to refold a road map is
differently.

—JONES'S, Rule of the Road

One of the most unusual and frustrating unsolved problems in
modern combinatorial theory, proposed many years ago by
Stanislaw M. Ulam, is the problem of determining the number
of different ways to fold a rectangular "map." The map is precreased along vertical and horizontal lines to form a matrix of
identical rectangles. The folds are confined to the creases, and
the final result must be a packet with any rectangle on top and
all the others under it. Since there are various ways to define
what is meant by a "different" fold, we make the definition
precise by assuming that the cells of the unfolded map are
numbered consecutively, left to right and top to bottom. We
wish to know how many permutations of these n cells, reading
from the top of the packet down, can be achieved by folding.
Cells are numbered the same on both sides, so that it does not
matter which side of a cell is "up" in the final packet. Either
end of the packet can be its "top," and as a result every fold
will produce two permutations, one the reverse of the other.
The shape of each rectangle is irrelevant because no fold can
rotate a cell 90 degrees. We can therefore assume without altering the problem that all the cells are identical squares.

The simplest case is the 1-by-n rectangle, or a single strip of
n squares. It is often referred to as the problem of folding a
strip of stamps along their perforated edges until all the
stamps are under one stamp. Even this special case is still un-

solved in the sense that no nonrecursive formula has been found for the number of possible permutations of n stamps. Recursive procedures (procedures that allow calculating the number of folds for n stamps provided that the number for $n-1$ stamps is known) are nonetheless known. The total number of permutations of n objects is $n!$ [that is, factorial n, or $n \times (n-1) \times (n-2) \ldots \times 1$]. All $n!$ permutations can be folded with a strip of two or three stamps, but for four stamps only 16 of the $4!=24$ permutations are obtainable [see Figure 27]. For five stamps the number of folds jumps to 50 and for six stamps it is 144. John E. Koehler wrote a computer program, reported in a 1968 paper, with which he went as high as $n=16$, for which 16,861,984 folds are possible. W. F. Lunnon, in another 1968 paper, carried his results to $n=24$, and in a later paper, to $n=28$. Koehler showed in his article that the number of possible stamp folds is the same as the number of ways of joining n dots on a circle by chords of two alternating colors in such a way that no chords of the same color intersect.

The simplest rectangle that is not a strip is the trivial 2-by-2 square. It is easy to find that only eight of the $4!=24$ permutations can be folded, half of which (as explained above) are

Figure 27

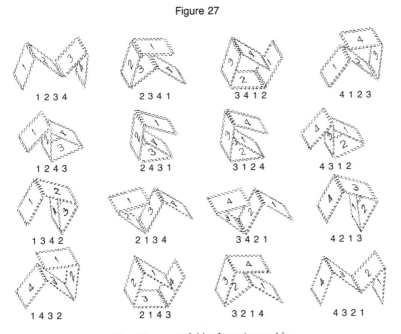

1 2 3 4   2 3 4 1   3 4 1 2   4 1 2 3

1 2 4 3   2 4 3 1   3 1 2 4   4 3 1 2

1 3 4 2   2 1 3 4   3 4 2 1   4 2 1 3

1 4 3 2   2 1 4 3   3 2 1 4   4 3 2 1

The 16 ways to fold a four-stamp strip

reversals of the other half. The 2-by-3 rectangle is no longer trivial, because now it becomes possible to tuck one or more cells into open pockets. This greatly confuses matters. As far as I know, nothing has been published on the nonstrip rectangles. I was able to fold 60 of the $6! = 720$ permutations (10 folds for each cell on top), but it is possible I missed a few.

An amusing pastime is to find six-letter words that can be put on the 2-by-3 map (lettering from left to right and from the top down) so that the map can be folded into a packet that spells, from the top down, an anagram of the original word. Each cell should be labeled the same on both sides to make it easier to identify in the packet. For example, it is not hard to fold ILL-FED to spell FILLED and SQUIRE to spell RISQUE. On the other hand, OSBERG (an anagram for the last name of the Argentine writer Jorge Luis Borges that appears on page 361 of Vladimir Nabokov's novel *Ada*) cannot be folded to BORGES, nor can BORGES be folded to OSBERG. Can the reader give a simple proof of both impossibilities?

The 2-by-4 rectangle is the basis of two map-fold puzzles by Henry Ernest Dudeney (see page 130 of his *536 Puzzles & Curious Problems*. Scribner's, 1967). Dudeney asserts there are 40 ways to fold this rectangle into a packet with cell No. 1 on top, and although he speaks tantalizingly of a "little law" he discovered for identifying certain possible folds, he offers no hint as to its nature. I have no notion how many of the $8! = 40,320$ permutations can be folded.

When one considers the 3-by-3, the smallest nontrivial square, the problem becomes fantastically complex. As far as I know, the number of possible folds (of the $9! = 362,880$ permutations) has not been calculated, although many paperfold puzzles have exploited this square. One was an advertising premium, printed in 1942 by a company in Mt. Vernon, N.Y., that is diagrammed in Figure 28. On one side of the paper there are the faces of Mussolini and Hitler. On the back of the remaining cell of the same row is the face of Tojo, the wartime prime minister of Japan. Above this cell is a prison window with open spaces die-cut between two bars; below the Tojo cell a similar window appears on the back of the cell, as indicated by the dotted lines. The problem is to fold the square into a packet so that at each end two of the faces appear behind the bars; that is, so that on each side of the packet the top cell bears a picture of a window and directly under it a face shows through the open slots between the bars. The fold is not difficult, but it does require a final tuck.

Figure 28

A map-fold problem from World War II

A much tougher puzzle using a square of the same size is the creation of Robert Edward Neale, a Protestant minister, professor of psychiatry and religion at the Union Theological Seminary and the author of the influential book *In Praise of Play* (Harper & Row, 1969). Neale is a man of many avocations. One of them is origami, the Oriental art of paper folding, a field in which he is recognized as one of the country's most creative experts. Magic is another of Neale's side interests; his famous trick of the bunny in the top hat, done with a folded dollar bill, is a favorite among magicians. The hat is held upside down. When its sides are squeezed, a rabbit's head pops up. (The interested reader can obtain *Bunny Bill*, a manuscript describing the fold, from Magic, Inc., 5082 North Lincoln Avenue, Chicago, Ill. 60625. The fold is far from simple, by the way.)

Figure 29 shows Neale's hitherto unpublished Beelzebub puzzle. Start by cutting a square from a sheet of paper or thin cardboard, crease it to make nine cells, then letter the cells (the same letter on opposite sides of each cell) as indicated. First try to fold the square into a packet that spells (from the top down) these eight pseudonyms of the fallen angel who, in Milton's *Paradise Lost,* is second in rank to Satan himself: Bel Zeebub, Bub Blezee, Ube Blezbe, Bub Zelbee, Bub Beelze, Zee Bubble, Buz Lebeeb, Zel Beebub. If you can master these names, you are ready to tackle the really fiendish one: Beelzebub, the true

Figure 29

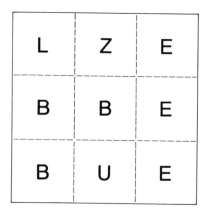

Robert Edward Neale's Beelzebub puzzle

name of "the prince of the devils" (Matthew 12:24). Its extremely difficult fold will be explained in the Answer Section. No one who succeeds in folding all nine names will wonder why the general map-folding problem is still unsolved.

Neale has invented a variety of remarkable paper-fold puzzles, but there is space for only two more. One is in effect a nonrectangular "map" with a crosscut at the center [*see Figure 30*]. The numbers may represent six colors: all the 1-cells are one color, the 2-cells a second color and so on. Here opposite sides of each cell are different. After numbering or coloring as shown at the top in the illustration, turn the sheet over (turn it *sideways*, exchanging left and right sides) and then number or color the back as shown at the bottom. The sheet must now be folded to form a curious species of tetraflexagon. (Tetraflexagons were the topic of an earlier column that is reprinted in *The 2nd Scientific American Book of Mathematical Puzzles & Diversions*.)

To fold the tetraflexagon, position the sheet as shown at the top in the illustration. (It helps if you first press the creases so that the solid lines are what origamians call "mountain folds" and the dotted lines are "valley folds.") Reach underneath and seize from below the two free corners of the 1-cells, holding the corner of the upper cell between the tip of your left thumb and index finger and the corner of the lower cell between the tip of your right thumb and finger. A beautiful maneuver can now be executed, one that is easy to do when you get the knack even though it is difficult to describe. Pull the corners simultaneously down and away from each other, turning each 1-cell

Figure 30

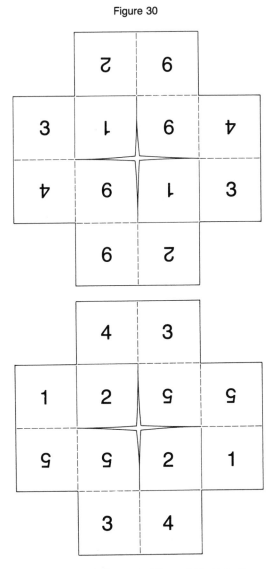

Front *(top)* and back *(bottom)* of the unfolded tetraflexagon

over so that it becomes a 5-cell as you look down at the sheet. The remaining cells will come together to form two opentop boxes with a 6-cell at the bottom of each box [*see Figure* 31].

Shift your grip to the two inside corners of the 5-cells—corners diagonally opposite the corners you were holding. Push down on these corners, at the same time pulling them apart. The boxes will collapse so that the sheet becomes a flat 2-by-2

Figure 31

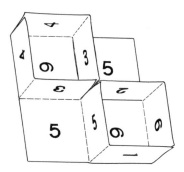

First step in folding the tetraflexagon

tetraflexagon with four 1-cells on top and four 2-cells on the underside [*see Figure* 32]. If the collapsing is not properly done, you will find a 4-cell in place of a 1-cell, and/or a 3-cell in place of a 2-cell. In either case simply tuck the wrong square out of sight, replacing it with the correct one.

Figure 32

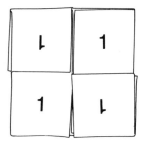

Final step in folding the tetraflexagon

The tetraflexagon is flexed by folding it in half (the two sides going back), then opening it at the center crease to discover a new "face," all of whose cells have the same number (or color). It is easy to flex and find faces 1, 2, 3 and 4. It is not so easy to find faces 5 and 6.

One of Neale's most elegant puzzles is his "Sheep and Goats," which begins with a strip of four squares and a tab for later gluing [*see Figure* 33.] Precrease the sheet (folding it both ways) along all dotted lines. Then color half of each square [*dark grey in illustration*] black—on both sides, as if the ink had soaked right through the paper.

The strip is folded as shown in steps *a*, *b*, *c* and *d*. The first

Figure 33

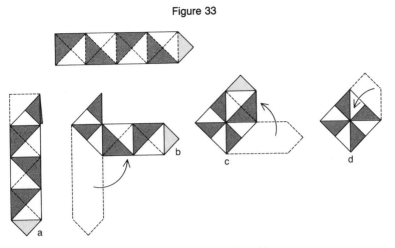

Neale's "Sheep and Goats" problem

fold is back and down. The next three folds are valley folds, first to the right, then up, then left. After the last fold slide the tab *under* the double-leaved black triangle at the top left of the square. Glue the tab to the bottom leaf of the triangle. You should now have a square with four black and four white triangles on each side. These are the sheep and goats.

The problem is: By folding only along precreased lines, change the paper to a square of the same size that is all white on one side and all black on the other. In other words, separate the sheep from the goats. It is not easy, but it is a delight to make the moves rapidly once you master the steps—which I shall diagram in the Answer Section, along with the answer to the tetraflexagon puzzle. (To make the manipulations smoother, it is a good plan to trim a tiny sliver from all single edges after the square has been folded and glued.)

Anyone interested in learning some of Neale's more traditional origami figures will find six of his best (including his Thurber dog) in Samuel Randlett's *The Best of Origami* (E. P. Dutton, 1963). Some of his dollar-bill folds (including the jumping frog) are in *Folding Money: Volume II*, edited by Randlett (Magic, Inc., 1968).

ANSWERS

A simple proof that on the two-by-three rectangle OSBERG cannot be folded to spell BORGES (or vice versa) is to note that in each case the fold requires that two pairs of cells touching only at their corners would have to be brought together in the final packet. It is evident that no fold can put a pair of such cells together.

The square puzzle with the faces and prison windows is solved from the starting position shown. Fold the top row back and down, the left column toward you and right, the bottom row back and up. Fold the right packet of three cells back and tuck it into the pocket. A face is now behind bars on each side of the final packet. The central face of the square cannot be put behind bars because its cell is diagonally adjacent to each of the window cells.

Space prevents my giving solutions for the eight pseudonyms of Beelzebub, but Beelzebub itself can be obtained as follows. Starting with the layout shown, fold the bottom row toward you and up to cover *BBE*. Fold the left column toward you and right to cover *ZU*. Fold the top row toward you and down, but reverse the crease between *L* and *Z* so that *LZ* goes between *B* and *B* on the left and the upper *E* goes on top of the lower *E*. You now have a rectangle of two squares. On the left, from the top down, the cells are *BLZBUB*, on the right *EEE*. The final move is difficult. Fold the right panel (*EEE*) toward you and left. The three *E*'s are tucked so that the middle *E* goes between *Z* and *B*, and the other two *E*'s together go between *B* and *L*. Once you grasp what is required it is easier to combine this awkward move with the previous one. The result is a tightly locked packet that spells Beelzebub. The solution is unique. If the cells of the original "map" are numbered 1 through 9, the final packet is 463129785.

To find the 5-face of the tetraflexagon, start with face 1 on the top and 2 on the bottom. Mountain-fold in half vertically, left and right panels going back, so that if you were to open the flexagon at the center crease you would see the 4-face. Instead of opening it, however, move left the lower inside square packet (with 4 and 3 on its outsides) and move right the upper square packet (also with 4 and 3 on its outsides). Insert your fingers and open the flexagon into a cubical tube open at the top and bottom. Collapse the tube the other way. This creates a new tetraflexagon structure that can be flexed to show faces 1, 3 and 5.

A similar maneuver creates a structure that shows faces 2, 4 and 6. Go back to the original structure that shows faces 1, 2, 3 and 4 and repeat the same moves as before except that you begin with the 2-face uppermost and the 1-face on the underside.

Figure 34 shows how to separate the sheep from the goats:
(1) Start with the two-color square folded as shown.
(2) Fold in half along the horizontal diagonal by folding the

Figure 34

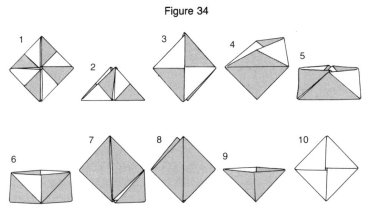

Solution to the "Sheep and Goats" problem

bottom corner up to make a "hat" with a white triangle at the lower left-hand corner.

(3) Open the hat's base and continue opening until you can flatten the hat to make the small square shown.

(4) Insert a left finger into the pocket on the right of the upper face of drawing No. 3. Pull upward and flatten as shown.

(5) Turn the paper over sideways and repeat the previous move on the other side. The result is a rectangle with a white triangle in the upper right-hand corner.

(6) Open the rectangle into a cubical tube open at the top and bottom. Collapse the tube the other way to make a rectangle again, except that now it is colored as shown.

(7) Insert your right thumb into the pocket on the left of drawing No. 6, lift up the flap and flatten it as shown.

(8) Turn the paper over sideways and do the same on the other side. You should now have a small square, black on both sides.

(9) Reach into the square from above, open it and flatten to make an inverted hat, black on both sides.

(10) Open the hat by separating its bottom points and flatten the large square that results. It will be the same size as the square you started with, but now it is all white on one side and all black on the other—all sheep and all goats.

Repeating the same sequence of moves will mix the sheep and the goats again. With practice the folds can be done so rapidly that you can hold the square out of sight under a table for just a few moments and produce the change almost as if by magic.

ADDENDUM

The "little law" that Henry Ernest Dudeney hinted about in connection with his map-fold problem has probably been rediscovered. Mark B. Wells of the Los Alamos Scientific Laboratory used a computer to confirm that the 2×3 map has 10 folds for each cell on top. The program also found that the order-3 square has 152 folds for each cell on top. In his 1971 paper Lunnon proved that for any rectangular map every cyclic permutation of every possible fold is also a possible fold. Thus it is necessary to determine only the folds for one cell on top because the cyclic permutations of these folds give all the other folds. For example, since 123654789 is a possible fold, so also are 236547891, 365478912 and so on. It is a strange law because the folds for cyclic permutations differ wildly. It is not yet known whether the law applies to all polyomino-shaped maps or to maps with equilateral triangles as cells.

In his 1971 paper Lunnon used an ingenious diagram based on two perpendicular slices through the center of the final packet. He was able to write a simple backtrack program for x-by-y maps, extend the problem to higher dimensions and discover several remarkable theorems. For example, the edges of one cross section always diagram x linear maps of y cells each, and the edges of the other cross section diagram y linear maps of x cells each.

The 2×3, 2×4, 2×5, 2×6, and 3×4 maps have respectively 60, 320, 1980, 10512, and 15552 folds. The order-3 square has 1368 folds, the order-4 has 300608, the order-5 has 186086600. In all cases the number of folds is the same for each cell on top, as required by cyclic law. The order-2 cube, folded through the fourth dimension, has 96 folds. The order-3 cube has 85109616. Many other results are tabulated by Lunnon in his 1971 paper, but a nonrecursive formula for even planar maps remains elusive.

The linear map-fold function, as Lunnon calls it, is the limit approached by the ratio between adjacent values of the number of possible folds for a $1 \times n$ strip. It is very close to 3.5. In his unpublished 1981 paper Lunnon narrows the upper and lower bounds to 3.3868 and 3.9821.

In 1981 Harmony Books in the United States, and Pan Books in England brought out a large paperback book called *Folding Frenzy*. It contains six 3×3 squares, with red and green patterns on both sides, and five pages partially die-cut. Without removing any pages, there are nine puzzles to solve by folding the squares. The puzzles are credited to Jeremy Cox.

In describing one of his map-fold puzzles (*Modern Puzzles,*

No. 214) Dudeney mentioned a curious property of map folds that is not at all obvious until you think about it carefully. It applies not only to rectangular maps, but also to maps in the shape of any polyomino; that is, a shape formed by joining unit squares at their edges. Assume that any such map is red on one side, white on the other. No matter how it is folded into a 1 × 1 packet, the colors on the tops of each cell will alternate regardless of which side of the packet is up. If the cells of the map are colored like a checkerboard, with each cell the same color on both sides, the final packet (after any sort of folding) will have leaves that alternate colors. If the checkerboard coloring is such that each cell is red on one side, white on the other, all cells in the folded packet will have their red sides facing one way, their white sides facing the other way.

It occurred to me in 1971 that the parity principles involved here could be the basis for a variety of magic tricks. One appeared under the title "Paradox Papers" in Karl Fulves' magic periodical, *The Pallbearers Review*. It goes like this: Fold a sheet of paper twice in each direction so that the creases make 16 cells. It is a good plan to fold the paper each way along every crease to make refolding easier later on.

Assume in your mind that the cells are checkerboard colored black and red, with red at the top left corner. Five red playing cards are taken from a deck and someone selects one of them. With a red pencil jot the names of the five cards in five cells, using abbreviations such as 4D and QH. Tell your audience that you are taking cells at random, but actually you must put the name of the chosen card on one of the "black" cells, and the other four names on "red" cells.

Have another card chosen, this time from a set of five black cards. Turn the sheet over, side for side, and jot the names of the five black cards on cells, again apparently at random. Use a black pencil. Put the chosen card on a "red" cell, the others on "black" cells.

Ask someone to fold the sheet any way he likes to make a 1 × 1 packet. With a pair of shears, trim around the four sides of the packet. Deal the 16 pieces on the table. Five names will be seen, all the same color except for one—the chosen card of the other color. Turn over the 16 pieces. The same will be true of the other sides.

Gene Nielsen, in the May 1972 issue of the same journal, suggested the following variant. Pencil X's and O's on all the cells, alternating them checkerboard fashion. Turn over the sheet horizontally, and put exactly the same pattern on the other side. Spectators will not realize that each cell has an

X on one side, a *O* on the other. Someone folds the sheet randomly into a packet. Pretend you are using PK to influence the folding so that it will produce a startling result. Trim the sides of the packet and spread the pieces on the table. All *X*'s face one way, all *O*'s face the other way.

Swami, a magic periodical published in Calcutta by Sam Dalal, printed my "Paper Fold Prediction" in its July 1973 issue. Start by numbering the cells of a 3×3 sheet from 1 through 9, taking the cells in the usual way from left to right and top down. Put the digits on one side of the paper only. After someone folds the sheet randomly, trim the sides of the packet and spread the pieces. Add all the numbers showing. Reverse the pieces and add the digits on the other sides. The two sums will be different. Explain that by randomly folding the sheet, the nine digits are randomly split into two sets. Clearly there is no way to know in advance what either sum will be when the pieces are spread.

Repeat the same procedure, but this time use a 4×4 square with cells numbered 1 through 16. The sheet is randomly folded and the edges trimmed. Before spreading the pieces, hold them to your forehead and announce that the sum will be 68. Put down the packet, either side uppermost, and spread the pieces. The numbers showing will total 68. Discard the pieces before anyone discovers that the sum on the reverse sides also is 68.

The trick works because if the original square has an odd number of cells, the sums on the two sides will not be equal. (On the 3×3 they will be 20 and 25.) However, if the square has an even number of cells, the sum is a constant equal to $(n^2 + n)/4$ where n is the highest number. You can now repeat the trick with a 5×5 square, but instead of predicting a sum, predict that the *difference* between the sums on the two sides of the pieces will be 13.

The principle applies to cells numbered with other sequences. For example, hand a wall calendar to someone and ask him to tear out the page for the month of his birth. He then cuts from the page any 4×4 square of numbers. The sheet is folded, the packet trimmed, the pieces spread, and the visible numbers added. The sum will be equal to four times the sum of the sheet's lowest and highest numbers. You can predict this as soon as you see the square that has been cut, or you can divine the number later by ESP.

Some other suggestions. Allow a spectator to write any digit he likes in each cell of a sheet of any size, writing left to right and top down. As he writes, keep a running total in your head by subtracting the second number from the first, adding the third, subtracting the fourth, and so on. The running total is likely to fluctuate between plus and minus. The number you end with, whether plus or minus, will be the difference between the two sums after the sheet is folded, trimmed, and the pieces spread.

Magic squares lend themselves to prediction tricks of a similar nature. For example, suppose a 4×4 map bears the numbers of a magic square. After folding, trim only on two opposite sides of the packet. This will produce four strips. Have someone select one of the four. The other three are destroyed. You can predict the sum of the numbers on the selected strip because it will be the magic square's constant. Of course you do not tell the audience that the numbers form a magic square.

Many of these tricks adapt easily to nonsquare sheets, such as a 3×4. The underlying principles deserve further exploration by mathematical magic buffs.

BIBLIOGRAPHY

Avec des nombres et des lignes (With Numbers and Lines). A. Sainte-Lague. Paris: Librarie Vuibert, 1937, third edition 1946. The linear $(1 \times n)$ postage stamp problem is discussed in Part 2, Chapter 2, pages 147–162. I am indebted to Victor Meally for telling me of this earliest known analysis of the problem.

"Contributions à l'étude du problème des timbres poste." J. Touchard. *Canadian Journal of Mathematics,* Vol. 2, 1950, pages 385–398.

"A Map-Folding Problem." W. F. Lunnon. *Mathematics of Computation,* Vol. 22, January 1968, pages 193–199.

"Folding a Strip of Stamps." John E. Koehler. *Journal of Combinatorial Theory,* Vol. 5, September 1968, pages 135–152.

"Multi-Dimensional Stamp Folding." W. F. Lunnon. *The Computer Journal,* Vol. 14, February 1971, pages 75–79.

"The Map-Folding Problem Revisited." W. F. Lunnon. *Mathematics of Computation,*

"Postage Stamps." Stuart Baker, in "Computer Solutions to Three Topological-Combinatorial Problems," 1976, unpublished paper.

"The Devil's Fold." Robert Neale. *Games,* November/December 1979, page 33.

"Bounds for the Map-Folding Function." W. F. Lunnon. Cardiff, August, 1981. Unpublished.

8

A SET OF QUICKIES

The following problems are of the "quickie" type in the sense that they are quickly stated and, at least so I believed when I first gave them, not hard to solve if properly approached. Some are joke questions, and others contain booby traps to catch the unwary.

Problem 1: You want to construct a rigid wire skeleton of a one-inch cube by using 12 one-inch wire segments for the cube's 12 edges. These you intend to solder together at the cube's eight corners.

"Why not cut down the number of soldering points," a friend suggests, "by using one or more longer wires that you can bend at sharp right angles at various corners?"

Adopting your friend's suggestion, what is the smallest number of corners where soldering will be necessary to make the cube's skeleton rigid? (Philip G. Smith, Jr.)

Problem 2: An intelligent horse learns arithmetic, algebra, geometry and trigonometry but is unable to understand the Cartesian coordinates of analytic geometry. What proverb does this suggest? (Howard W. Eves, in *Mathematical Circles*, Vol. 1.)

Problem 3: Your king is on a corner cell of a chessboard and your opponent's knight is on the corner cell diagonally opposite. No other pieces are on the board. The knight moves first. For how many moves can you avoid being checked? (From David L. Silverman's collection of game problems, *Your Move*.)

Problem 4: Nine heart cards from an ordinary deck are arranged [*see Figure* 35] to form a magic square so that each row, column and main diagonal has the largest possible constant sum, 27. (Jacks count 11, queens 12, kings 13.) Drop the requirement that each value must be different. Allowing duplicate values, what is the largest constant sum for an order-3 magic square that can be formed with nine cards taken from a deck? (M. G.)

Figure 35

A magic square with nine hearts

Problem 5: Make a statement about n that is true for, and only true for, all values of n less than one million. (Leo Moser.)

Problem 6: Why would a barber in Geneva rather cut the hair of two Frenchmen than of one German?

Problem 7: With a black pencil draw a closed curve of any shape you please. With a red pencil draw a second curve of the same kind on top of the first one, never passing through a previously created intersection. Circle all points where one curve crosses the other [*see Figure* 36]. Prove that the number of such points is even. (M. G.)

Problem 8: Place a familiar mathematical symbol between 2 and 3 to express a number greater than 2 and less than 3.

Figure 36

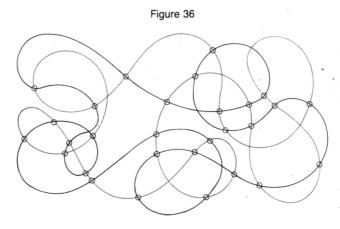

A topological theorem

Problem 9: A six-story house (not counting the basement) has stairs of the same length from floor to floor. How many times as high is a climb from the first to the sixth floor as a climb from the first to the third floor?

Problem 10: Each of the two equal sides of an isosceles triangle is one unit long. Without using calculus, find the length of the third side that maximizes the triangle's area.

Problem 11: What three positive integers have a sum equal to their product?

Problem 12: A string, lying on the floor in the pattern shown in Figure 37, is too far away for you to see how it crosses itself at points *A, B* and *C*. What is the probability that the string is knotted? (L. H. Longley-Cook, *Fun with Brain Puzzlers.*)

Problem 13: If *AB, BC, CD* and *DE* are common English words, what familiar word is DCABE? (David L. Silverman, *Word Ways*, August 1969.)

Problem 14: *Time,* March 7, 1938, reported that one Samuel Isaac Krieger claimed to have found a counterexample to Fermat's unproved last theorem. Krieger announced that it was $1,324^n + 731^n = 1,961^n$, where *n* is a certain positive integer greater than 2, and which Krieger refused to disclose. A reporter on *The New York Times,* said *Time,* easily proved that Krieger was mistaken. How?

Figure 37

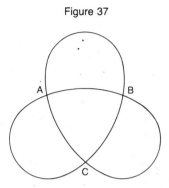

Is the string probably knotted?

Problem 15: What familiar English word begins and ends with *und*?

Problem 16: A man arrives at a random spot several miles from the Pentagon. He looks at the building through binoculars. What is the probability that he will see three of its sides? (F. T. Leahy, Jr.)

Problem 17: Change 11030 to a person by adding two straight line segments.

Problem 18: A boy and a girl are sitting on the front steps of their commune.
   "I'm a boy," said the one with black hair.
   "I'm a girl," said the one with red hair.
   If at least one of them is lying, who is which? (Adapted from a problem by Martin Hollis, in *Tantalizers*.)

Problem 19: A "superqueen" is a chess queen that also moves like a knight. Place four superqueens on a five-by-five board so that no piece attacks another. If you solve this, try arranging 10 superqueens on a 10-by-10 board so that no piece attacks another. Both solutions are unique if rotations and reflections are ignored. (Hilario Fernandez Long.)

Problem 20:

      A B C D
      D C B A
      • • • •
      ‾‾‾‾‾‾‾
      1 2 3 0 0

ABCD are four consecutive digits in increasing order. *DCBA* are the same four in decreasing order. The four dots represent the same four digits in an unknown order. If the sum is 12,300, what number is represented by the four dots? (W. T. Williams and G. H. Savage, *The Strand Problems Book.*)

Problem 21: A "primeval snake" is formed by writing the positive integers consecutively along a snaky path [*see Figure* 38]. If continued upward to infinity, every prime number will fall on the same diagonal line. Explain. (M. G.)

Figure 38

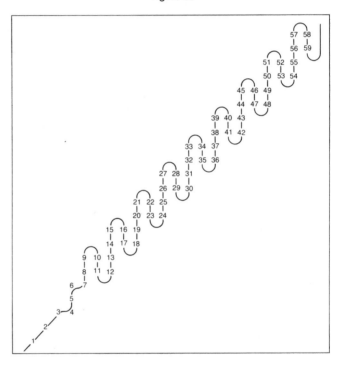

The primeval snake

Problem 22: Find two positive integers, *x* and *y*, such that the product of their greatest common divisor and their lowest common multiple is *xy*.

Problem 23: "Feemster owns more than a thousand books," said Albert.

"He does not," said George. "He owns fewer than that."

"Surely he owns at least one book," said Henrietta.

If only one statement is true, how many books does Feemster own?

Problem 24: In this country a date such as July 4, 1971, is often written 7/4/71, but in other countries the month is given second and the same date is written 4/7/71. If you do not know which system is being used, how many dates in a year are ambiguous in this two-slash notation? (David L. Silverman.)

Problem 25: Why are manhole covers circular instead of square?

Problem 26: How many different 10-digit numbers, such as 7,829,034,651, can be written by using all 10 digits? Numbers starting with zero are excluded.

Problem 27: Many years ago, on a sultry July night in Omaha, it was raining heavily at midnight. Is it possible that 72 hours later the weather in Omaha was sunny?

Problem 28: What well-known quotation is expressed by this statement in symbolic logic?

$$2B \lor \sim 2B = ?$$

Problem 29: Regular hexagons are inscribed in and circumscribed outside a circle [*see Figure* 39]. If the smaller hexagon has an area of three square units, what is the area of the larger hexagon? (Charles W. Trigg, *Mathematical Quickies.*)

Figure 39

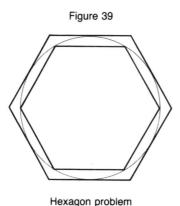

Hexagon problem

Problem 30: "I was n years old in the year n^2," said Smith in 1971. When was he born?

Problem 31: If you think of any base greater than 2 for a number system, I can immediately write down the base without asking you a question. How can I do this? (Fred Schuh, *The Master Book of Mathematical Recreation*.)

Problem 32: What was the name of the Secretary General of the United Nations 35 years ago?

Problem 33: You have one red cube and a supply of white cubes all the same size as the red one. What is the largest number of white cubes that can be placed so that they all abut the red cube, that is, a positive-area portion of a face of each white cube is pressed flat against a positive-area portion of a face of the red cube. Touching at corner points or along edges does not count. (M. G.)

Problem 34: What four consecutive letters of the alphabet can be arranged to spell a familiar four-letter word? (Murray R. Pearce, *Word Ways*, February 1971.)

Problem 35: Figure 40 is a diagram of a deep circular lake, 300 yards in diameter, with a small island at the center. The two black spots are trees. A man who cannot swim has a rope a few yards longer than 300 yards. How does he use it as a means of getting to the island?

Figure 40

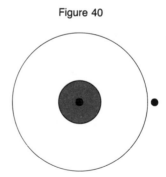

Lake, island, and trees

Problem 36: A boy, a girl and a dog are at the same spot on a straight road. The boy and the girl walk forward—the boy at four miles per hour, the girl at three miles per hour. As they proceed the dog trots back and forth between them at 10 miles per hour. Assume that each reversal of its direction is instantaneous. An hour later, where is the dog and which way is it facing? (A. K. Austin.)

ANSWERS

1. The smallest number of soldering points remains eight no matter how wires are bent. Because an odd number of edges meet at each corner of a cube, every point requires soldering.

2. Do not put Descartes before the horse.

3. You can evade check forever. Head toward the board's center, always moving your king to a color opposite to that of the knight. Since a knight changes the color of its cell at every move, whenever the king is on a color different from the knight's, no knight move can check the king. Your only danger lies in being trapped in a corner where you can be forced to move diagonally and be checked by the knight's next move.

4. The highest constant is 36 [*see Figure* 41].

5. One answer: "The value of n is less than one million."

6. He makes twice as much money.

Figure 41

Answer to the card problem

7. The black curve divides the plane into a number of regions. Trace a round trip along the red curve and it is obvious that every region you enter you must also leave or you will never get back to where you started. Since each entrance and exit is a pair of crossing points, the total number of such spots must be even.

Figure 42

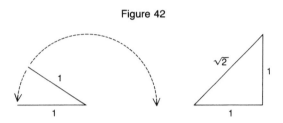

Answer to the triangle problem

8. 2.3.

9. Two and a half times as high.

10. With one unit side as base and the other unit side free to rotate [*see Figure* 42], the triangle's area is greatest when the altitude is maximum. The third side will then be the square root of 2.

11. $1 + 2 + 3 = 1 \times 2 \times 3$.

12. Only two of the eight possible combinations of crossings create a knot, making the probability of a knot $2/8 = 1/4$.

13. House.

14. The first number, 1,324, raised to any power must end in 6 or 4. The other two numbers, 731 and 1,961, raised to any power must end in 1. Since no number ending in 6 or 4, added to a number ending in 1, can produce a number ending in 1, the equation has no solution.

15. Underground.

16. One proof that the probability is 1/2: Suppose the man has a *Doppelgänger* directly opposite him on the other side of the Pentagon's center and the same distance away. If either man sees three sides, his double must see only two. Since there is an equal probability that either man is at either spot, the probability is 1/2 that he will see three sides.

17. HOBO.

18. The four possible true-false combinations for the two statements are *TT, TF, FT* and *FF*. The first is eliminated because we were told that one statement is false. The second and third are eliminated because in each case, if one person lied, the other cannot have spoken truly. Therefore both lied. The boy has red hair, the girl black hair.

19. Figure 43 shows the two solutions.

Figure 43

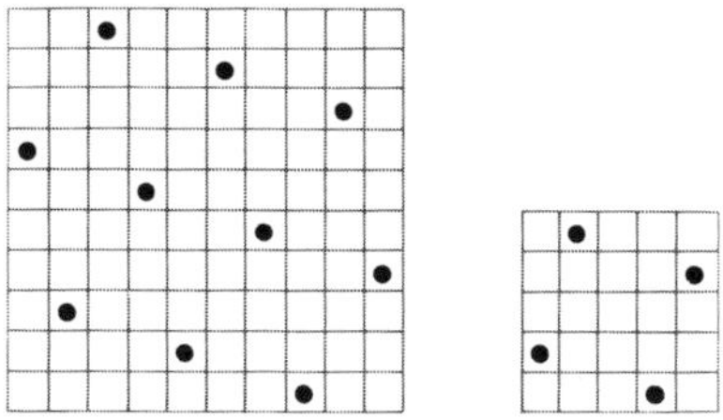

Superqueen solutions

20. If $ABCD = 1,234$, it is impossible to obtain a sum as large as 12,300. If $ABCD = 3,456$, it is impossible to obtain a sum as small as 12,300. Therefore $ABCD = 2,345$, from which it is easy to determine that the four dots stand for 4,523.

21. It is well known that every prime greater than 3 is one more or one less than a multiple of 6. It is easy to see that every number of the form $6n \pm 1$ must fall on the same diagonal, therefore the diagonal is certain to catch every prime.

22. Any two positive integers.

23. There are three permissible combinations of true and false for the three statements: *TFF, FTF* and *FFT*. The only noncontradictory combination is *FTF*, which means that Feemster owns no books at all.

24. Each month has 11 ambiguous dates (a date such as 8/8/71 is not ambiguous), making 132 in all.

25. A square manhole cover, turned on edge, could slip through its hole and fall into the sewer.

26. Ten digits can be permuted in $10! = 3,628,800$ different ways. A 10-digit number cannot start with zero, so that we must subtract $3,628,800/10 = 362,880$ to obtain the answer: 3,265,920.

27. No, because after 72 hours it would have been midnight again.

28. "To be or not to be, that is the question."

Figure 44

Calculating hexagon areas

29. Instead of inscribing the hexagon as shown, turn it to the position shown in Figure 44. The grey lines divide the larger hexagon into 24 congruent triangles, 18 of which form the smaller hexagon. The ratio of areas is 18 : 24 = 3 : 4, and so if the smaller hexagon has an area of three, the larger one has an area of four.

30. Smith was born in 1892. He was 44 in $44^2 = 1936$.

31. I write "10." This is *any* base written in that base system's notation.

32. The same as it is now.

33. Twenty white cubes can abut the grey cube. Arrange seven white cubes as shown in Figure 45. The grey cube goes on top as indicated. Seven more white cubes, in the same pattern and position as the first layer, form layer No. 3. In between layer No. 1 and layer No. 3 six more white cubes can be placed: two on each of two opposite sides of the grey cube and single cubes on the remaining two sides.

Figure 45

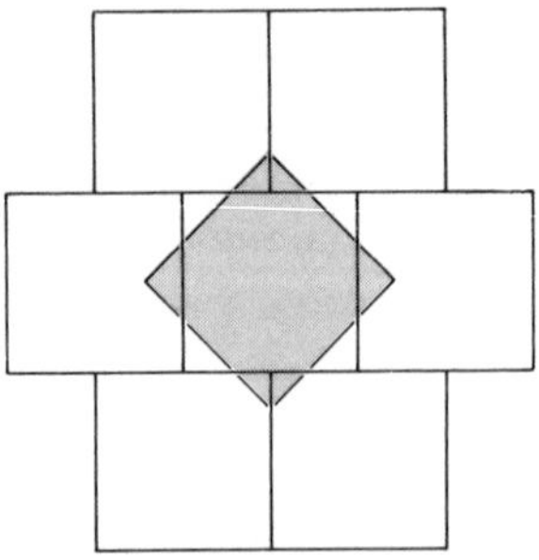

Arrangement of the cubes

34. The consecutive letters *RSTU* will spell "rust" or "ruts."

35. He ties one end of the rope to the tree at the edge of the lake, walks around the lake holding the other end of the rope and ties that end to the same tree. The doubled rope is now firmly stretched between the two trees, making it easy for the man to pull himself through the water, by means of the rope, to the island.

36. The dog can be at any point between the boy and the girl, facing either way. Proof: At the end of one hour, place the dog anywhere between the boy and the girl, facing in either direction. Time-reverse all motions and the three will return at the same instant to the starting point.

ADDENDUM

The answers to the 36 "quickie" type problems brought more surprises by mail than any previous collection of short problems. Readers caught ambiguous phrasings, indulged in amusing quibbles, found alternate and sometimes better answers, spotted some errors, and argued that the last problem is meaningless. I shall comment on this correspondence, taking the problems in numerical order, and add some further observations of my own.

(4) C. C. Cousins, Charles W. Bostick, and others noticed that four of the court cards in the illustration for the answer to this problem are incorrectly drawn. The Jack of Diamonds and the Jack of Clubs should be one-eyed, and the King of Spades and King of Hearts should face the other way. Some readers thought the Queen of Diamonds should face the other way. But Bostick took the trouble to examine 30 different decks made in the United States and found that in 18 of them the Queen of Diamonds faced right, and in 12 cases she faced left, so this card cannot be considered wrong.

(5) This theorem is related to a paradox of induction that I came across in Karl Popper's *Conjectures and Refutations* where he attributes it to J. Agassi. "All events occur before the year 3000." Since this statement has so far been confirmed by every event in the history of the universe, some theories of induction are forced to regard it as strongly confirmed, thus suggesting that it is highly probable the world will end before 3000.

(8) Larry S. Liebovitch, instead of using a decimal point, solved this problem by using "ln," the symbol of "natural log of." Thus $2 \ln 3 = 2.19 +$.

(11) Problem E 2262, in *The American Mathematical Monthly* (November 1971, pages 1021–1023), By G. J. Simmons and D. E. Rawlinson, generalized this question by asking for all other sets of k positive integers of which the same statement could be made. It turns out it can be made for all positive integers, but only a very small set have unique answers. When $k=2$, the only answer is $2+2=2\times2$. Our problem provided the only answer for $k=3$. For $k=4$ it is $2+4+1+1=2\times4\times1\times1$.

Readers of the periodical showed that for all values of k not exceeding 1,000, the only values with unique solutions are 2, 3, 4, 6, 24, 114, 174, and 444. It is possible, the editor comments, that no other values have unique answers other than the eight listed.

(12) James A. Ulrich was the first to argue that the probability of the string's being knotted is 1 because there is no way a closed loop of string can exist without its ends being tied.

(13) "House" remains the best answer, but less familiar words such as "ye" and "el" allow other solutions. George A. Miller sent a computer printout of 269 alphabetized answers, and all the words (from "abhor" to "wavey") are found in standard dictionaries.

(14) Martin Kruskal provided a photocopy of the *New York Times* account (February 22, 1938) of Samuel Isaac Krieger's preposterous claim to have disproved Fermat's last theorem. He had saved the clipping since he had seen it as a small boy.

(15) Solomon W. Golomb proposed "underfund" and "underwound" as alternate answers.

(16) The probability of 1/2 that a distant viewer will see three sides of the Pentagon is correct only as a limit as the viewer's distance from the Pentagon building approaches infinity. My solution ignored the fact that there are five infinitely long strips, each crossing the building, inside of which both the viewer and his *Doppelgänger* can see only two sides (and if very close to the Pentagon, only one side). This was pointed out by readers too numerous to list.

The probability is zero, commented P. H. Lyons, "if the smog in Washington is anything like what it is here in Toronto."

(17) Walter C. Eberlin and David Dunlap independently added two strokes to 11030 so that when it is viewed in a mirror it spells "peon," a word closely related to "hobo" in meaning.

Richard Ellingson took advantage of the fact that I did not

specify that the lines of 11030 could not be rearranged. His solution was:

(20) Hans Marbet, of Switzerland, pointed out that if *ABCD*
are replaced by any four consecutive digits, and the four dots
replaced by the same digits in the order *CDAB,* there is a valid
solution for a number system with the base $A+B+D$. For example, in base 16:

$$4567$$
$$7654$$
$$\underline{6745}$$
$$12300$$

(21) A. P. Evans, William B. Friedman, and others wrote to
say that you don't need to know that all primes greater than 3
have the form of $6n$ plus or minus 1. Only four parallel diagonals can be drawn through the snake. Call them, top to bottom, *A,B,C,D*. All numbers on *A* are divisible by 3 and therefore cannot be prime. All on *B* and *D* are divisible by 2, and
hence cannot be prime. Therefore all the primes must fall on
C. "I don't suppose it would be sporting," Evans adds, "to ask
readers to come up with a diagram on which a straight line can
be drawn that contacts all prime numbers and *only* prime
numbers."

(23) When George said that Feemster "owns fewer than
that," I meant him to mean fewer than the amount specified by
Albert. If "that" is taken to refer to "a thousand" instead of
"more than a thousand," however (as many readers pointed
out), Feemster could own exactly 1,000 books as well as none.

(24) Howard J. Frohlich passed along a friend's view that a
date such as 8/8/71 could also be called "ambiguous" because
you do not know whether the first 8 refers to the day or the
month.

(25) Since I failed to ask why manhole covers *and holes* are
round instead of square, scores of readers sensibly replied that
the covers are round to fit the holes. John W. Stack cited as his
authority for this answer M. A. Nhole's *Comprehensive Review of
Equilateral Rectangular Beams and Circular Receptacles,* pages 31–
4207, published in 1872 by the Sewer and Street Company,
Inc. P. H. Lyons had another answer: To reduce the decisions
a sewer worker has to make in replacing the cover.

Some covers and holes *are* square, according to John Bush, who told of a recent explosion near his home in Brooklyn that blew off a Consolidated Edison square manhole cover. After the smoke cleared the cover was found at the bottom of the manhole. *"Geometria invincibilis est,"* Bush concluded.

(28) The Hamlet rebus, "To be or not to be," was invented by Golomb, a fact I did not know when I gave it.

Jim Levy wrote to say that strictly speaking the symbol for "or" should be one that represented exclusive disjunction (either but not both) rather than inclusive disjunction (either or both), otherwise the statement implies that a person can be and not be simultaneously.

(32) "Can you answer this?" Golomb wrote in 1971. "No, U Thant!"

(33) The problem of the touching cubes was one I thought of several years ago and had answered with 20 cubes. I was staggered to receive two different solutions, each with 22 cubes. Figure 46 shows how five white cubes can abut the top side of the grey cube. Since none extends beyond line *AB*, this formation can go on four sides of the grey cube [*see Figure* 47].

Figure 46

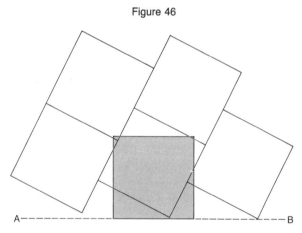

Five cubes abut one side of the shaded cube

Two more cubes plug up the holes on face A and its opposite side. This solution was first received from Kenneth J. Fawcett, Jr., and later from Rudolf K. M. Bergan, Michael J. and Alice E. Fischer, Leigh Janes, K. B. Mallory, Allen J. Schwenk and George Starbuck.

Figure 47

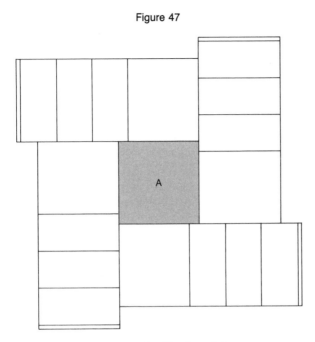

Arrangement for 22-cube solution

The other solution, found independently by Bergan, Rudolph A. Krutar and Robert S. Holmes, is shown as drawn by Holmes [*Figure* 48]. Eight cubes go on two opposite faces of the grey cube, and six abut the grey cube in the middle layer. Even the fact that as many as eight nonintersecting unit squares can overlap one unit square is, as far as I know, a previously unknown result.

Stanley Ogilvy later pointed out that because the bottom corners of the three lowest squares in Figure 46 are not on a horizontal line, there is just enough room below them to permit three more squares, joined face to face, to go beneath the other five squares. This provides another way for eight cubes to abut one face, and leads to another solution with 22 cubes.

While I was still recovering from the 22-cube solution, Holmes (who is working for his doctorate in particle physics at the University of Rochester) delivered the knockout punch: a 24-cube solution! Later Janes, in collaboration with Michael Bradley, reported a 23-cube solution.

It is hard to believe, but as far as I know no one has seriously considered before the simple question of how many unit cubes

Figure 48

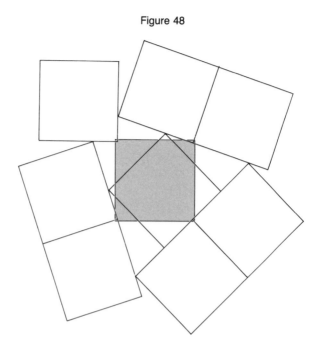

Another 22-cube solution

can share a positive surface area with a given unit cube. My innocent answer of 20 is indeed the best if one adds the condition that the surface of the given unit cube (we shall distinguish it from the others by making it grey and leaving the other cubes white) be completely covered by the touching cubes. This proviso, however, was not part of the original problem.

Holmes's technique begins with placing seven white cubes on one grey face [*see Figure* 49]. Three pairs of cubes (think of each pair as being glued together) are placed around a face of the grey cube so that the midpoint of each pair touches a corner of the grey face. A seventh white cube (*P*, drawn with broken lines) overlaps the grey face as indicated, the two faces having an axis of symmetry shown as a diagonal line. By rotating the white cubes clockwise, keeping corner A on the left edge of the fixed grey face and preserving the bilateral symmetry, the pattern shown in Figure 50 is reached. If the broken-line cube *P* is now moved up a trifle, the two meeting corners of each pair of glued cubes can have a tiny positive-area overlap with a corner of the grey face. These three overlaps can be made arbitrarily small without allowing the white cubes

Figure 49

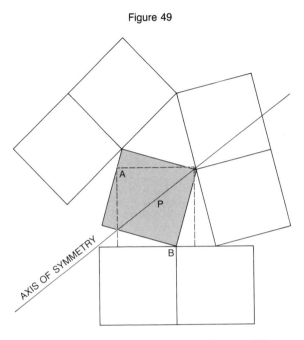

Beginning arrangement for Holmes's 24-cube solution

Figure 50

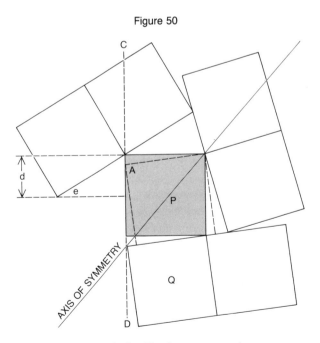

Step 2 in the 24-cube arrangement

P and *Q* to project to the left beyond the vertical line *CD*. As a result angle *e* and distance *d* can also be made arbitrarily small.

This pattern of seven cubes goes on the front and rear faces of the grey cube. Then one cube goes exactly on top of the grey cube, another goes flush against the grey cube's base and two more cubes abut the right face of the grey cube. Although 18 cubes now abut five faces of the grey one, its sixth face (on the left) remains completely exposed. Figure 51 shows the grey cube with the exposed face toward you. On both its left and right sides are seven cubes; they are not shown in the drawing. (Also not shown are the single cubes above and below and the two cubes that abut the grey cube's back face.) Cube *K* is placed so that it overlaps the top of the grey face along a thin horizontal strip of height *d* that can be arbitrarily small. This allows five more cubes to abut the face, below cube *K*, as shown, bringing the total number of touching cubes to 24 ($7 + 7 + 1 + 1 + 2 + 1 + 5$).

Figure 51

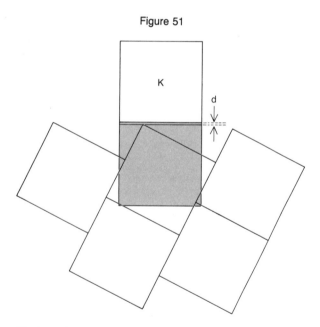

Final step in 24-cube solution covers remaining exposed face

A full proof of this construction would be long and tedious, but interested readers should have no difficulty convincing themselves that it can be done in spite of the extremely minute overlaps that are involved. The 24-cube solution is probably maximum, although proving it appears to be formidable. Until that is done there will remain the gnawing suspicion that one or more additional white cubes can somehow be squeezed in.

Theodore Katsanis posed an interesting related problem. What is the *minimum* number of unit cubes that can touch another unit cube in such a way that no other cube can be added? If "touch" is defined as in the original problem, the answer is obviously six. Suppose, however, we enlarge the meaning of touch to include contact along edges and corners. The maximum problem is again trivial, answered by 26 cubes, but the minimum problem is not. The lowest Katsanis could get is nine, but perhaps some reader can do better.

(34) Commented P. H. Lyons: "I hope some readers tried other languages, such as Hawaiian. If the letters of the alphabet need not be in alphabetical order, I have a fair-sized list of other answers in English."

(35) William B. Friedman proposed placing the rope so that half of it is high above the water and the other half still higher. The man could then walk along the lower rope, holding on to the upper one, and not even get wet. If the rope is hemp, wrote P. H. Lyons, the man could smoke it and fly to the island.

(36) This question (about the boy, the girl and the dog) stirred up a hornet's nest. Some mathematicians defended the answer as being valid, others insisted the problem has *no* answer because it is logically contradictory. There is no way the three can start moving, it was argued, because the instant they do the dog will no longer be between the boy and the girl. This plunges us into deep waters off the coast of Zeno. The issue is discussed in detail in Chapter 13.

9

TICKTACKTOE GAMES

"It's as simple as tit-tat-toe, three-in-a-row,
and as easy as playing hooky. I should
hope we can find a way that's a little more
complicated than *that*, Huck Finn."

—MARK TWAIN,
The Adventures of Huckleberry Finn

Ticktacktoe (the spelling varies widely) is not nearly so simple
as Tom Sawyer thought. When Charles Sanders Peirce wrote
his *Elements of Mathematics,* a textbook that was not published
until 1976, he included a 17-page analysis of only the side
opening of this ancient game. It was one of Peirce's many an-
ticipations of "modern math." Today's progressive teachers
frequently use ticktaktoe to introduce their pupils to such con-
cepts as the intersection of sets, rotational and mirror-reflec-
tion symmetry, and higher Euclidean space. In this chapter we
consider some unusual aspects of the game not covered in two
earlier columns reprinted in *The Scientific American Book of
Mathematical Puzzles & Diversions* (Chapter 4), and *Mathematical
Carnival* (Chapter 16).

The traditional game, as most readers surely know, is a draw
if both players do their best. From time to time pictures of a
ticktacktoe game appear in advertisements and cartoons, and
sometimes they provide pleasant little puzzles. For example, on
May 13, 1956, in the New York *Herald Tribune,* there was an
IBM advertisement with the unfinished game at the left in Figure 52. Which player went first, assuming that the players were
not stupid? It takes only a moment to realize that *O* could not

Figure 52

	X	
	O	X
		O

X	O	O
O	X	O
X	O	X

X		X
	O	
O		X

Three easy ticktacktoe puzzles

have gone first or *X* would have played the top left corner on his second move. The other two patterns are almost as trivial. Does the center one, from a Little Lulu cartoon in *The Saturday Evening Post* (January 16, 1937), depict a possible game? At the right is a pattern from an advertisement by publisher Lyle Stuart in *The New York Times* (June 1, 1971). In which cell must the last move have been made?

If the first player, say *X*, opens in the center cell, he can force a draw that always ends with the same basic pattern. This underlies several prediction tricks. For example, the magician draws the finish of a game, with all cells filled, on a square sheet of paper that he turns face down without letting anyone see it. He then plays a ticktacktoe game with someone, writing on another square sheet. After the game ends in a draw he turns over his "prediction." The two patterns match cell for cell.

The technique is explained in Figure 53. *X* plays the center opening. If *O* marks any corner cell, *X* forces the draw shown at the left in the illustration (moves are numbered in order of play). It is only necessary for *X* to remember where to make his second move, since all moves are forced from then on; a simple rule for the second move is to consider the corner opposite *O*'s first move and then play adjacent to it on the clockwise side. If *O* responds to the opening with a side cell, *X* forces the draw shown at the right. In this situation only *O*'s moves are forced

Figure 53

O_2	O_4	X_5
X_7	X_1	O_8
O_6	X_3	X_9

O_8	O_2	X_3
X_5	X_1	O_6
O_4	X_9	X_7

A ticktacktoe prediction trick

and X must remember how to play his next four moves. The following simple rule was proposed by a magician who signed himself "Thorson" when he described this trick in the September 1960 issue of *M.U.M.*, official organ of the Society of American Magicians: X makes his second, third and fourth moves adjacent and clockwise to O's previous moves, and his fifth move in the only remaining empty cell.

Note that the two final results are identical. Of course, each game can be played in any of four different orientations. The magician, recalling which corner of his inverted prediction has the O surrounded by three X's, casually turns over the square sheet along the proper axis—orthogonal or diagonal—so that his predicted pattern matches the orientation of the game just completed.

The trick can even be repeated. This time X substitutes counterclockwise for clockwise in the rules, having drawn a prediction that is a mirror image of the preceding one. The two predictions will not match in any orientation and few people will realize that they are mirror reflections of each other.

Dozens of variations of planar ticktacktoe have been analyzed. Standard games on squares of higher order than 3, when the goal on an order-n board is to get n in a row, are uninteresting because the second player can easily force a draw. My first column on ticktacktoe discussed games in which counters are moved over the board (one such version goes back to ancient Greece), and toetacktick, in which the first to get three in a row *loses*.

A. K. Austin's "wild ticktacktoe," in which players may use either X or O on every move, was shown to be a first-player win in my *Sixth Book of Mathematical Games*, Chapter 12, Problem 3. What about "wild toetacktick," in which players can choose either mark on each move and the first three-in-a-row loses? In 1964 Solomon W. Golomb and Robert Abbott independently found that the simple symmetry strategy by which the first player can force at least a draw in standard toetacktick also applies to the wild version. A center opening is followed by playing directly opposite the other player's moves, always choosing X if he played O and O if he played X. The question remains: Does the first player have a *winning* strategy in wild toetacktick? Abbott made an exhaustive tree diagram of all possible plays and proved that the second player too can force a draw. Tame toetacktick also is a draw if both sides play rationally.

An amusing variation appears in David L. Silverman's book of game puzzles, *Your Move*. The rules are the same as in standard ticktacktoe except that one player tries to achieve a draw and the other player wins if either of them gets three in a row. Can the reader show that no matter who plays first the player trying to force a row of three can always do so? Silverman does not answer this in his book, but I shall give his solution in the Answer Section.

It is impossible to describe all the other planar variants that have been proposed, such as using numbers or letters as marks for the goal of forming a certain sum or spelling a certain word; playing on the vertexes of curious nine-point graphs (for a game on one such graph see my *Mathematical Magic Show*, Chapter 5, Problem 5); using counters with X on one side and O on the other, with the counters turned over according to specified rules. Games have been marketed in which flip-overs are randomized by concealed magnets that may or may not reverse a counter or by tossing beanbags at a board to cause cubical cells to alter their top symbols by rotating.

If ticktacktoe is played on an unlimited checkerboard, it is a trivial win for the first player if the goal is to get four or any smaller number of one's marks in an orthogonal or diagonal row. When the goal is five in a row, this game is far from trivial. It is the ancient Oriental game known as go-moku (five stones) in Japan, where it is played on the intersections of a go board. (The game is sold in the U.S. by Parker Brothers under the name of Pegity.) Although it is widely believed that a first-player winning strategy exists, this has not yet, to my knowledge, been proved.

There is no doubt about the first player's strong advantage in unrestricted go-moku. Indeed, it is so overwhelming that in Japan the standard practice is to weaken the first player by not allowing the following moves:

(1) A move that simultaneously creates a "fork" of two or more intersecting rows of open threes. By "open three" is meant any pattern in which a play will form a row of four adjacent stones that is open at both ends. There is one exception. A fork move is permitted if it is the only way to block an opponent from completing a row of five.

(2) A move that forms a row of more than five. In other words, the winning move must be *exactly* five.

In master play, both rules are usually applied only to the first player. Under these rules, the game is commonly called "renju" in Japan.

It has been conjectured that if there is a winning strategy for the first player in unrestricted go-moku, on a large enough board, there will be a winning strategy on a sufficiently large board even if the prohibitions are observed, but this is far from established. Even if a winning strategy is found for unrestricted go-moku, difficult questions will remain. What is the *smallest* board on which the first player can win? What is the *shortest* way to win? The two questions may or may not be answered by the same line of play.

It is not possible that the second player has a winning strategy in unrestricted go-moku or similar games in any dimension. The bare bones of the simple *reductio ad absurdum* proof, first formulated by John Nash for the game of hex, are as follows. Assume that a second-player winning strategy exists. If it does, the first player can make an irrelevant, random first play—a play that can only be an asset—and then, since he is now in effect the second player, win by appropriating the second player's strategy. Because this contradicts the assumption, it follows that no second-player winning strategy exists. The first player can either win or at least force a draw if the game allows draws.

Go-moku is a stimulating game. To catch its special flavor the reader is urged to study a position from Silverman's book [*see Figure* 54] and determine how O can play and win in five moves. Note that X has an open-end diagonal of three, which he threatens to lengthen to an open-end row of four.

Figure 54

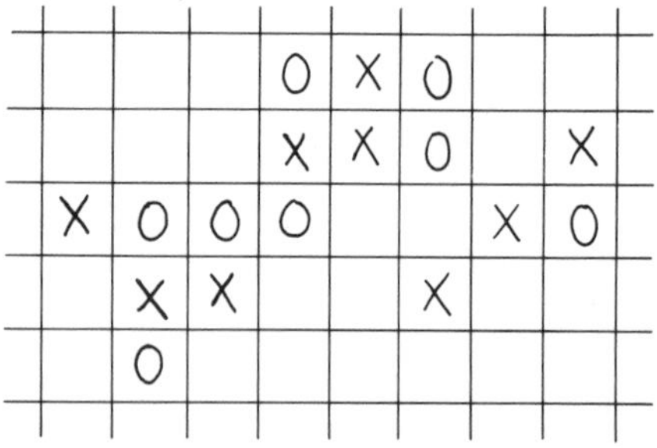

Go-moku problem: 0 to play and win

When ticktacktoe is extended to three dimensions, the first player wins easily on an order-3 cube by first taking the center cell. As Silverman points out, if the first player fails to open with the center cell, the second player can win by taking it; if the center is permanently prohibited to both players, the first player has an easy win. Three-dimensional toetacktick (the first row of three loses) is also a win for the first player. He plays the same strategy used for forcing a draw in planar toetacktick: He first takes the center and then always plays symmetrically opposite his opponent. Since drawn positions are impossible on the order-3 cube, this technique forces the second player eventually to form a row of three. Daniel I. A. Cohen, in a paper listed in the bibliography, proves that, as in the case of planar toetacktick, this is a unique winning strategy. The first player loses if he does not open by taking the central cell, and also loses if, after making this first move, he does not follow antipodal play.

Draw games *are* possible on the order-4 cube, but whether the first player can force a win is not, as far as I know, positively established. (There cannot, of course, be a second-player win because of Nash's proof.) As with go-moku, the first player has a strong advantage and a winning strategy is believed to exist. Many computer programs for this game have been written, but the complexity of play is so enormous that I do not think a first-player win has yet been rigorously demonstrated. About a dozen readers have sent me what they consider winning strategies, but detailed formal proofs are still unverified. Most of the strategies involve first taking four of the eight central cells and then proceeding to a forced win. Virtually nothing is known about three-dimensional games where counters are allowed to move from cell to cell.

Another unexplored type of 3-space game is one in which two players alternately draw from a limited supply of unit cubes of two or more colors to build a larger cube with some winning objective in view, for example, using cubes of n colors and trying to get a row, on an order-n cube, in which all n colors appear. For such games gravity imposes restraints, since cubes cannot be suspended in midair.

Because drawn games of standard ticktacktoe are possible in 2-space on an order-3 board, and in 3-space on an order-4 board, it was once conjectured that in a space of n dimensions the smallest board allowing a draw was one with $n+1$ cells on a side. It turned out, however, that although in n-space a board of order $n+1$ or higher always allows a draw, it is sometimes possible for an n-space board of fewer than $n+1$ cells on a side to allow a draw. This was first established about 1960 by Alfred W. Hales, when he constructed a draw on the order-4 hypercube, or fourth-dimension cube.

Several readers have sent informal but probably valid proofs that the first player can always win on the order-4 hypercube. Whether or not he can force a win on the order-5 hypercube is yet another of the many unanswered questions about extensions and variants of what most people, like Tom Sawyer, regard as a "simple" game.

ANSWERS

The second game in Figure 52 is not possible. Zero must have played first and last, but X had a win before the final move, so the last move would not have been made. In the third game, X could have completed a win if his first two moves had been on either side, therefore the first two moves must have been diagonally opposite, and his final move in the top right corner.

These two problems are so easily solved that I will add here a difficult one that involves what chess players call retrograde analysis. Figure 55 shows the pattern after two perfect players have agreed to a draw. Your task is to determine the first and last moves. If you can't solve it, you will find the solution in the *Journal of Recreational Mathematics*, Vol. 11, No. 1, 1978, page 70. The problem had been earlier posed in the same journal by Les Marvin.

Figure 55

What were the first and last moves?

In Silverman's first problem, X can always win, regardless of whether he plays first or second. Assume that the cells are numbered (left to right, top to bottom) from 1 to 9. Here is Silverman's proof:

If X begins, he takes 1. O must take 5, otherwise X can get three of his marks in a row by standard ticktacktoe strategy. $X2$ forces $O3$, then $X4$ forces $O7$, which completes three O's in a line, giving X the win.

If O starts the game, he has a choice of corner, side or center opening. If he opens at the center (5), X responds with 1. If the move is $O2$, $X7$ forces $O4$, then $X9$ forces $O8$, which loses. If O's second move is 3, $X4$ forces $O7$, which also loses. If O's second move is 6, $X7$ forces O to lose at 4. If O's second move is 9, X2 forces O3, then X4 forces O to lose at 7. All other lines of play are symmetrically equivalent.

If O opens at the side, say at 4, X5 will win. As before, there are four basically different continuing lines of play: (1) O1, X3, O7 (loses), (2) O2, X3, O7, X9, O1 (loses), (3) O3, X9, O1, X8, O2 (loses), (4) O6, X3, O7, X9, O1 (loses).

A corner opening by O, say at 1, is met with X5, which leads again to four basically different continuations: (1) O2, X7, O3 (loses), (2) O3, X8, O2 (loses), (3) O6, X8, O2, X7, O3 (loses), (4) O9, X2, O8, X3, O7 (loses).

When this game is played on a four-by-four field (X winning if there are four of either mark in a row, O winning if the final position is drawn), the play is so enormously more complex, Silverman informs me, that it has not yet been fully analyzed.

O wins Silverman's go-moku problem by playing O1 [see Figure 56]. X2 is forced, O3 forces X4, O5 forces X6, then O7 creates an open-end diagonal row of four O's, which X cannot block. If X plays at either end, O wins by playing at the opposite end. As Silverman points out in his book, O wins only by counterattacking. He loses quickly if he plays defensively by trying to block X's open-end diagonal row of three.

Figure 56

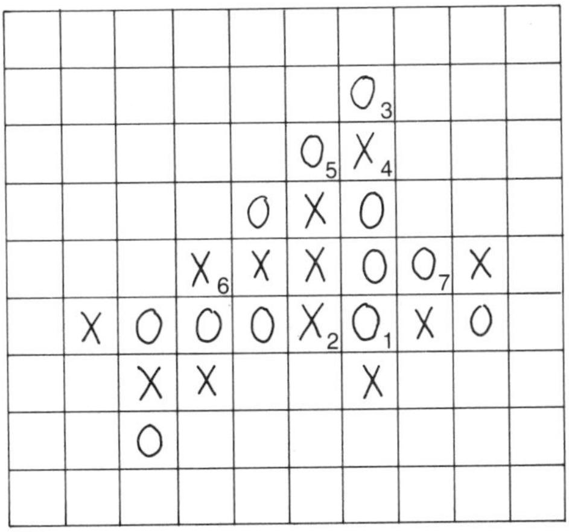

Solution to the go-moku problem

Note that when X plays on the cell marked 2 it creates a fork. This is permitted, however, because the move is forced. It is the only way to prevent O from winning on the next move.

ADDENDUM

John Selfridge reports that a solution has been found for his "4× infinity" ticktacktoe. This is played on a strip that is four cells high and infinitely wide, the winner being the first to get four of his marks in an orthogonal or diagonal row. Carlyle Lustenberger, in his master's thesis in computer science at Pennsylvania State University, developed a computer program with a winning strategy for the first player on a four-by-30 board. The actual lower bound for the width is a few cells shorter, but I have not obtained the details.

The three-by-infinity board is a trivial win for the first player on his third move; indeed, the same win can be achieved if only one cell is added to the side or corner cell of the traditional order-3 ticktacktoe field. The five-by-infinity board remains unsolved. If a win for the first player could be found on this board, it would, of course, solve the go-moku game when it is played on an arbitrarily large square, with no restrictive rules.

Owen Patashnik, of Bell Laboratories, was the first to write a computer program that establishes a first-player win in 4× 4×4 ticktacktoe. I had the honor of announcing the verification of his 1977 program in my *Scientific American* column of January 1979. It required 1,500 hours of computing time, and has been likened to the computer proof of the four-color map theorem in its length and complexity. I will say no more about it here because Patashnik has so thoroughly and amusingly reported on it in his paper listed in the bibliography. The program's set of 2,929 strategic moves for winning is probably far from minimal, but I know of no program that has reduced them.

In 1973 the Netherlands issued a 30 + 10 cents stamp depicting a drawn pattern in a ticktacktoe game.

Shein Wang, a computer scientist at the University of Guelph, Guelph, Ontario, Canada, has been publishing a monthly *Gomoku Newsletter* since 1979, and the university has, since 1975, been sponsoring a North American computer go-moku tournament. The programs have been steadily improving.

A popular variation of go-moku is on sale in the United States under the trade name *Pente*. Invented by Gary Gabel, it combines go-moku with elements of go. (See *Newsweek*, May 10, 1982, page 78.)

Several readers wrote to emphasize that Nash's proof applies only to unrestricted go-moku. The proof rests on the irrelevance of an extra stone, but in restricted go-moku the rules permit situations in which an extra stone can damage the player who owns it.

Henry Pollak and Claude Shannon apparently were the first to prove that the second player can force a draw in unrestricted n-in-a-row ticktacktoe on a large enough board when $n = 9$ or more. Their 1955 proof has not been published. It is given by T. G. L. Zetters in his answer to a problem, *American Mathematical Monthly,* Vol. 87, August–September 1980, pages 575–576. Zetters goes on to show how the proof can be extended to $n = 8$ or more. So far as I know, the question is still open for $n = 5$, 6 and 7.

W. F. Lunnon, writing in 1971 from University College, in Cardiff, gave a simple pairing strategy of unknown origin that guarantees a draw for the second player in 5×5 ticktacktoe. Number the cells as shown in Figure 57. Whenever the first player occupies a numbered cell, the second player takes the other cell of the same number. Since every line of five has a pair of like-numbered cells, the first player cannot occupy all five. If the first player takes the unlabeled center, the second player may take any cell, and if the cell he is required to take by the pairing strategy is occupied, he may play anywhere.

Lunnon also reported that he and Neil Sloane, of Bell Labs, had together found a remarkable second-player drawing strategy, based on cell pairing, for the 6×6 board. Not only does it

Figure 57

2	10	5	5	1
6	9	12	8	9
6	11		11	4
7	10	12	7	4
1	3	3	8	2

W. F. Lunnon's pairing strategy

block wins on any row, column or main diagonal, it also blocks a win on any broken diagonal! The cells are numbered as shown in Figure 58. As before, the strategy is to take the cell with the same number as the cell just taken.

Figure 58

1	13	2	13	3	12
6	14	5	14	4	12
7	8	15	9	10	15
16	3	11	1	16	2
17	4	11	6	17	5
7	8	18	9	10	18

Lunnon-Sloane second-player drawing strategy

There is more. The Lunnon-Sloane numbering leads to an elegant proof that 9-in-a-row unrestricted go-moku is a draw. Cover the infinite board with copies of the 6×6 matrix. The second player can force a draw by always taking the nearest cell with the same number as that of the previous play. It is easy to see that the first player can obtain no line longer than 8.

For $n \times n$ boards, n equal to 6 or higher, it is trivially easy to put a unique pair of numbers in each row of n cells and so provide a drawing strategy for the second player. For n equal to 3 or 4, no such labeling is possible, and the draw has to be established in uglier ways.

BIBLIOGRAPHY

Planar ticktacktoe and variants:

"The Tit-Tat-Toe Trick." Martin Gardner, in *Mathematics, Magic and Mystery*, Dover, 1956, pages 28–32.

"On Regularity and Positional Games." A. W. Hales and R. I. Jewett. *Transactions of the American Mathematical Society*, Vol. 106, 1963, pages 222–229.

The Theory of Gambling and Statistical Logic. Richard A. Epstein. Academic Press, 1967, pages 359–363.

"The Game of Noughts and Crosses." Fred Schuh in *The Master Book of Mathematical Recreations*, Dover, 1968, Chapter 3.

Your Move. David L. Silverman. McGraw-Hill, 1971, pages 69–78.

"The Solution of a Simple Game." Daniel I. A. Cohen. *Mathematics Magazine*, Vol. 45, September–October 1972, pages 213–216.

The New Elements of Mathematics. Charles Peirce. Edited by Carolyn Eisele. Humanities Press, 1976, Vol. 2, pages 11–24.

"Lines and Squares." Elwyn Berlekamp, John Conway, and Richard Guy in *Winning Ways*, Academic Press, 1982, Vol. 2, Chapter 22.

"Variations on a Game." J. Beck and L. Csirmaz. *Journal of Combinatorial Theory*, Series A, Vol. 33, November 1982, pages 297–315.

Go-moku:

Go and Go-moku. Edward Lasker. Dover, 1960. (Reprint of a 1934 book.)

"Experiments with a Learning Component in a Go-moku Playing Program." E. W. Elcock and A. M. Murray in *Machine Intelligence I*, edited by N. L. Collins and D. Michie. Oliver and Boyd, 1967.

"Automatic Description and Recognition of Board Patterns in Go-moku." A. M. Murray and E. W. Elcock in *Machine Intelligence II*, edited by Ella Dale and D. Michie. Oliver and Boyd, 1968.

"Renju." David Pritchard. *Games and Puzzles*, No. 76, Spring 1980, pages 26–28.

Computer programming:

"Computer Strategies for the Game of Cubic." W. G. Daly. M.S. thesis, MIT, 1961.

"The Concept of Strategy and Its Applications to Three Dimensional Tic-Tac-Toe." R. L. Citrenbaum. Systems Research Center, Report SRC-72-A-65-26. Case Institute of Technology, 1965.

"Efficient Representations of Optimal Strategy for a Class of Games." R. L. Citrenbaum. Systems Research Center, Report SRC-69-5. Case Western Reserve University, 1969.

"Strategic Pattern Generation: A Solution Technique for a Class of Games." R. L. Citrenbaum. Systems Development Corporation, Report SP-3505, Santa Monica, California, 1970.

"Machine Learning of Games." Ranan B. Banerji. *Computers and Automation*, November 1970, December 1970.

"Qubic: $4 \times 4 \times 4$ Tic-Tac-Toe." Owen Patashnik. *Mathematics Magazine*, Vol. 53, September 1980, pages 202–216.

10

PLAITING POLYHEDRONS

In Plato's dialogue *Phaedo,* Socrates tells a story in which the earth, viewed from outer space, appears "many-colored like the balls that are made of 12 pieces of leather." Historians take this to mean that the Greeks made balls by stitching together 12 leather pentagons stained with different colors and stuffing the interior to make the surface spherical. Rigid pentagons that are regular and identical would of course make a regular dodecahedron, one of the five Platonic solids.

There are all kinds of methods for constructing the five regular convex solids out of flat pieces of heavy paper or cardboard, and many problems have been proposed about ways of coloring their faces. The idea of weaving or braiding a regular solid from strips of paper seems to have been explored first by an English physician, John Gorham, who published in London in 1888 a now rare book about it: *A System for the Construction of Plaited Crystal Models on the Type of the Ordinary Plait.* His techniques were improved by A. R. Pargeter and by James Brunton in papers listed in the Bibliography. This year Jean J. Pedersen, a mathematics teacher at the University of Santa Clara, hit on an ingenious variation of the plaiting technique. It applies not only to the Platonic solids but also to many other polyhedrons, providing models of stunning multicolored symmetry and suggesting fascinating combinatorial theorems and puzzles.

Unlike Mrs. Pedersen's predecessors, who used crooked and asymmetrical basic patterns, she weaves each Platonic solid from *n* congruent straight strips. Assume that each strip is a different color and that each model has the following properties:

(1) Every edge is crossed at least once by a strip; that is, no edge is an open slot.

(2) Every color has an equal area exposed on the model's surface. (An equal number of faces will be the same color on all Platonic solids except the dodecahedron, which has bicolored faces when braided by this technique.)

Mrs. Pedersen has proved that if the above two properties are met, the number of necessary and sufficient bands for the tetrahedron, the cube, the octahedron, the icosahedron and the dodecahedron are respectively two, three, four, five and six.

Let us see how this works for the tetrahedron. Although the model can be plaited with one straight band, it will have some open edges. Therefore at least two bands are necessary. As shown in Figure 59, valley-crease each strip along the broken lines. (Scoring the lines with a hard pencil will facilitate clean folding.) Overlap two triangles as shown and fold the underneath strip into a tetrahedron. Wrap the other strip around two faces of this tetrahedron, then tuck the end triangle into the open slot. If you use construction paper of good quality and strips of different colors, the result is a rigid tetrahedron with two adjacent faces (of course, any two of its faces must be adjacent) of one color and two of the other color.

Figure 59

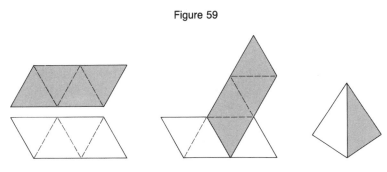

Plaiting a tetrahedron

To construct the cube three strips, each a different color, will do the trick [see Figure 60]. Valley-fold each along the broken lines. The reader can have the pleasure of weaving the three strips—it is quite easy—into a rigid cube. He will find that there are two essentially different ways to make a model with two faces of each color.

One method makes a cube that has adjacent pairs of faces with like colors. If you think of each band as being glued together where its end squares overlap, this model consists of three closed bands, each pair of which is linked. Imagine that

Figure 60

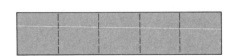
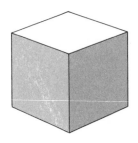

Plaited three-strip cube

the surface is flexible and that the cube is stuffed, like the leather dodecahedron mentioned by Plato, until it is spherical. The coloring, as Piet Hein has suggested, is a striking three-dimensional analogue of the familiar yin-yang symbol of the Orient. Like the yin-yang, it is asymmetrical (has either left or right handedness). Piet Hein proposes calling the three regions yin, yang and lee, the last two terms honoring C. N. Yang and T. D. Lee, the two Chinese-American physicists who shared a Nobel prize in 1957 for their role in overthrowing the symmetry law of parity.

The other way of plaiting the three strips produces a cube with opposite faces of like color. Again regard the three bands as being joined at their ends. Inspection reveals an unexpected structure. As Mrs. Pedersen has noted, the bands are topologically equivalent to the Borromean rings that are used as a trademark for Ballantine beer. Although the three bands cannot be separated, no pair is interlocked. If any one band is removed, the other two will slide apart.

The octahedron requires four valley-creased strips, each like the strip shown in Figure 61. These cannot be woven to make a model with opposite faces of the same color. (Can you prove it?) A model is possible, however, with like colors on pairs of adjacent faces, the four colors meeting at one corner and the same four, in reverse cyclic order, meeting at the diametrically opposite corner. A good procedure is to start with the two pairs of overlapping strips held together by a paper clip as shown in the illustration. Fold one pair into an octahedron, then weave the other pair around it, with both of the free ends tucked in slots, to achieve the desired color pattern. After the model is completed you can reach into the interior and remove the paper clips.

The octahedron is more difficult to make than the cube, but if the reader will set himself the task, he will find, as with all

Figure 61

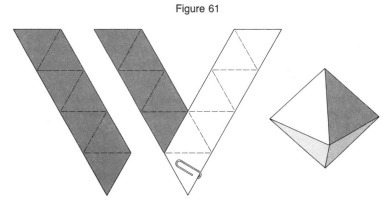

How strips are clipped together to weave an octahedron

such models, that there is an aesthetic delight in feeling the solid acquire permanent rigidity when the final tuck is made. Mrs. Pedersen has found that handsome models of this solid and the other four solids can be made by using colored cloth tape glued to construction paper for rigidity.

The icosahedron is woven with five valley-creased strips [*see Figure* 62]. A charming model can be constructed with each color on two pairs of adjacent faces, the pairs diametrically opposite each other. All five colors go in one direction around one corner and in the opposite direction, in the same order, around the diametrically opposite corner. Each band circles an "equator" of the icosahedron, its two end triangles closing the band by overlapping. In making the model, when the five overlapping pairs of ends surround a corner, all except the last pair can be pasted together or held with paper clips, which are re-

Figure 62

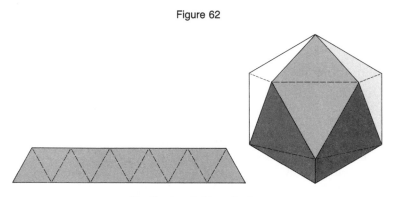

The five-banded icosahedron

moved later. The last overlapping end then slides into the proper slot. An expert will soon dispense with the use of paste or paper clips for this model.

Only the dodecahedron cannot be plaited with straight-sided strips so that each face is a solid color. Mrs. Pedersen discovered, however, that by using six strips the dodecahedron shown in Figure 63 can be woven. The obtuse angles made by the valley folds [*broken lines*] with the strip's sides are each 108 degrees, the interior angle of the regular pentagon. The broken lines must equal the shorter line segments on the sides, making each section of the strip a truncated pentagon.

Figure 63

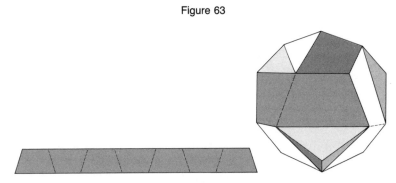

The six-banded dodecahedron

To construct the dodecahedron, the most difficult of the Platonic solids, Mrs. Pedersen suggests starting with three pairs of strips, each overlapped and glued together to make the curved, bracelet-like structure shown in Figure 64. Using two bracelets, overlap and glue together the pairs of ends to make a pair of braided closed bands. Slip one bracelet inside the other so that each circles a different equator of the dodecahedron. The third bracelet then is woven around a third equator, and its four free ends are tucked into slots on opposite sides of a pair of adjacent pentagonal faces. The technique is similar to the one used for making the cube with opposite faces of like color. Once the construction is mastered it is possible to use only paper clips to keep each bracelet together. The paper clips can be removed after the model is completed.

Note that every face of the finished dodecahedron has two colors. The same two colors are on the diametrically opposite face but are reversed in their arrangement. All diametrically opposite corners are mirror images in the order of the three or four colors that surround them. The model in the illustration,

Figure 64

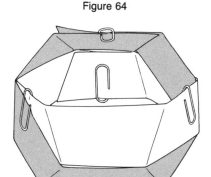

How pairs of strips are clipped together to weave
a dodecahedron

on which like colors are indicated by the same shade, appears
asymmetrical, but when the actual model is turned in one's
hands, its curious symmetry becomes apparent. The eight corners that are surrounded by exactly three colors mark the vertexes of an inscribed cube. The four corners surrounded by
three triangles mark the vertexes of an inscribed tetrahedron.

It is difficult to explain the exact procedure for plaiting the
last two models, so that I shall leave their construction as additional exercises for the patient and intrigued reader. It may
help to construct each solid first by conventional means, then
weave the required strips around it. I can only promise to report later if and where Mrs. Pedersen publishes instructions
for the Platonic solids as well as for more elaborate and less
regular polyhedrons that can also be formed by weaving congruent strips.

Mrs. Pedersen has devised a technique, involving the use of
gummed tape or adding-machine tape, for folding the strips
for all five models without drawing any fold lines. This technique, along with instructions for making what she calls a
golden dodecahedron (each face has a pentagonal hole surrounded by five triangles of different colors), are given in her
Fibonacci Quarterly article listed in the bibliography.

For years I was puzzled by the fact that Plato, repeating the
earlier views of Pythagoras and his followers, identified the
universe with the dodecahedron rather than the icosahedron,
which I took for granted to be more nearly spherical. I found
the answer recently in Volume I of Howard Eves's entertaining
work *In Mathematical Circles*. Contrary to almost everyone's intuition, it is the dodecahedron that is most like a sphere. If the

two solids are inscribed in the same unit sphere (a sphere with a radius of 1), the 20-faced icosahedron has a volume of 2.536 +, whereas the 12-faced dodecahedron has a *larger* volume of 2.785 +. Their surface areas are in the same ratio as their volumes: 9.574 + for the icosahedron, 10.514 + for the dodecahedron. The ancient Greeks had good reason to use the dodecahedron for their leather spheres.

If a cube and an octahedron are inscribed in a unit sphere, the cube has the greater volume and greater surface, and again their surface areas are in the same ratio as their volumes. The octahedron's volume and area are respectively 1.333 + and 6.928 +; the cube's volume and area, 1.539 + and 8. An interesting mechanical question, difficult to formulate precisely and perhaps even more difficult to answer, is which solid of each pair—cube or octahedron, icosahedron or dodecahedron—rolls more easily when used as a gaming device?

If a cube and an octahedron are inscribed in the same sphere, which solid surrounds the larger inscribed sphere? The surprising answer, as Eves explains, is that the two inner spheres are the same. This is also true of the inscribed spheres of a dodecahedron and an icosahedron that are inscribed in the same outer sphere.

Here are three polyhedron problems:

(1) What is the simplest *nonconvex* polyhedron that, like the cube, has a surface of n faces, each a unit square?

(2) If each face of a regular tetrahedron is a different color, how many different tetrahedrons can you make by using the same four colors? Rotations, of course, are not counted as different. Can you devise a simple formula that applies to all the Platonic solids, giving the number of different colorings possible when each of the n faces has a different color and the same n colors are used?

(3) If three colors are applied to a cube, each color going on two faces as in Mrs. Pedersen's plaited model, how many different colorings are possible? Again, as customary, rotations are not considered different. How many such cubes can be woven with Mrs. Pedersen's three bands, assuming there are no loose end squares that are not tucked in?

ANSWERS

1. The simplest nonconvex polyhedron with unit-square faces is the 30-face solid formed by attaching a unit cube to each face of a unit cube. Mrs. Pedersen found a way to braid this solid with three strips, each crossing once diagonally over every face of the solid. An infinite family of nonconvex polyhedrons with congruent square faces is obtained by joining any number of these crosses to form a chain.

2. A regular tetrahedron can be colored with four colors only in two ways, each a mirror reflection of the other. The simple formula that applies to all five Platonic solids is to divide the factorial of the number of faces by twice the number of edges. For example, the cube can be colored with six colors in $6!/24 = 30$ ways, the octahedron with eight colors in $8!/24 = 168$ ways, and so on.

3. A cube can be colored with three colors, each color going on two faces, in six ways: One with all pairs of opposite faces alike, two ways that are mirror images with all like colors on adjacent pairs of faces, and three ways with just one pair of opposite faces alike. Only the first three ways can be plaited with three five-square straight strips in the manner explained.

ADDENDUM

I was all wet in my argument that the dodecahedron is more spherelike than the icosahedron. Physicist F. C. Frank was the first to inform me that although the dodecahedron is closer in both volume and surface area to a sphere in which both are *inscribed,* the icosahedron is closer in both volume and area to a sphere that the two Platonic solids *circumscribe.* If you stuff each solid until it expands to make a sphere, you need less stuffing (in proportion to volume) to make the dodecahedron spherical. But if you carve away portions of each solid until you have a sphere, you carve away a smaller proportion of the volume of the icosahedron. Thus with respect to the insphere and circumsphere there is a standoff concerning which is the most spherical.

However, as Frank, Gary Goodman, Tom McCormick, Robert Dewar, and others pointed out, the sphere is well known for its property of having the greatest volume per surface area of any other solid. If *this* property is taken as the essence of sphericity, the icosahedron comes out ahead. There is nothing deceptive about our intuition when we observe the five Platonic solids and conclude that the tetrahedron looks the least like a sphere and the icosahedron looks the most like a sphere.

A definitive paper on the question, "Platonic Sphericity," by Norman T. Gridgeman, of Ottawa, was published in the *Journal of Recreational Mathematics* in 1973. Gridgeman upholds the commonsense view that the icosahedron is the most spherical, then goes on to discuss less obvious ways to measure "sphericity." He thinks Plato could have been influenced by knowing that the dodecahedron is closer to the circumsphere, and that this may have been augmented by the fact that the dodecahedron's pentagonal faces are closer to circles than the triangular faces of the icosahedron. Perhaps Plato was also influenced, Gridgeman speculates, by the correspondence between the dodecahedron's 12 sides and the 12 signs of the zodiac.

BIBLIOGRAPHY

"Plaited Polyhedra." A. R. Pargeter. *The Mathematical Gazette*, Vol. 43, May 1959, pages 88–101.

"The Plaited Dodecahedron." James Brunton. *The Mathematical Gazette*, Vol. 44, February 1960, pages 12–14.

Polyhedron Models. Magnus J. Wenninger. Cambridge University Press, 1971.

"Asymptotic Euclidean Type Constructions without Euclidean Tools." Jean J. Pedersen. *Fibonacci Quarterly*, Vol. 9, April 1971, pages 199–216.

"Some Whimsical Geometry." Jean J. Pedersen. *Mathematics Teacher*, Vol. 65, October 1972, pages 513–521.

Geometric Playthings. Jean J. Pedersen and Kent A. Pedersen. Troubadour Press, 1973.

"Geometric Sphericity." Norman T. Gridgeman. *Journal of Recreational Mathematics*, Vol. 6, Summer 1973, pages 106–210.

11

THE GAME OF HALMA

"An admirable place for playing halma," said
Chelifer, as they entered the Teatro
Metastasio.

—ALDOUS HUXLEY, *Those Barren Leaves*

Two new families of puzzles based on a long-neglected
counter-moving game have recently come to light. Each family
offers a series of unsolved problems and the opportunity to devise ingenious proofs that some solutions are impossible. The
puzzles stem from *Dialogue on Puzzles,* a splendid collection of
unusual problems by Kobon Fujimura and Michio Matsuda
published in 1971 in Japan. (Unfortunately the book is not
available in English.) Fujimura has translated the puzzle books
of Sam Loyd and Henry Ernest Dudeney into Japanese and is
the author of several delightful books that contain his own
original puzzles. The two new counter-moving puzzles are derived from one problem created by Matsuda.

Matsuda's problem exploits the simple rules of a popular
late-19th-century British proprietary game called Halma, after
the Greek word for leap. The game was invented in 1883 by
George Howard Monks, a 30-year-old Harvard Medical School
graduate who was then pursuing advanced studies in London.
He later became a prominent Boston surgeon. Halma is still
played in Britain but, although it was issued here in 1938 by
Parker Brothers, it has never caught on in this country.

The traditional Halma board has 16 cells on a side [*see Figure*
65]. If two players are competing, each begins by placing his
19 counters in a section called a "yard." There are two yards,

Figure 65

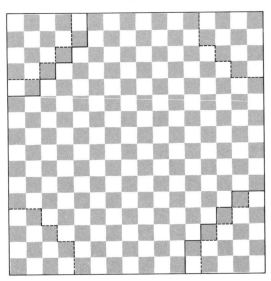

The Halma board

one at the top left corner of the board and the other at the bottom right corner. The counters are identical except that the two sets are of contrasting colors. The goal is to occupy the opposing player's yard, and the first player to move all his counters into the opposite yard is the winner. Two kinds of moves are allowed:

(1) A "step." This is a move, like the move of a chess king, to any one of the eight adjoining cells.

(2) A "hop." This is a leap over another counter, as in checkers, except that the leap may be made in any direction, orthogonal or diagonal. The jumped piece is not removed.

A connected chain of hops counts as a single move. It is not compulsory to make a hop. A player may continue a chain of jumps as long as possible or stop wherever he pleases. The color of a jumped piece does not matter; a chain of jumps may be a mixture of friendly and enemy counters. Steps and hops may not, however, be combined in the same move. Players alternate turns, moving one counter at a time.

Halma can also be played by four people, with each player having 13 counters. The yards are at the four corners of the board behind the boundaries indicated by the broken lines in the illustration. The four-player game can be each man for himself, with each seeking to reach the diagonally opposite

yard, or pairs of opposite (or adjacent) players can be partners
who help each other, and the first pair to yard all 26 of their
counters is the winner. Halma strategy is so complex, however,
that the game is best when only two people play.

Of many later games based on Halma the two most popular
in the United States have been Camelot and Chinese checkers,
both of which appeared on the market in the 1930's. Camelot
was a revival (with minor changes) by Parker Brothers of a late-
19th-century Parker game called Chivalry. Chinese checkers,
which has no connection whatever with China, is played on a
hexagonal-cell board that is usually shaped like a six-pointed
star. The hexagonal tessellation allows steps and hops in only
six directions. A French version of Halma, known as Grasshopper, can be played on a standard checkerboard [*see Figure* 66].
It is an excellent game.

Figure 66

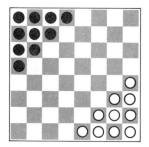

Grasshopper

To prevent a stubborn player in games of the Halma type
from forcing a draw by keeping a man permanently in his own
yard it is wise to add extra rules. Sidney Sackson, the New
York City game inventor and game collector, suggests the
following. If a counter can leave its own yard by jumping an
enemy counter, or by a chain of jumps that starts with a leap
over an enemy counter, it must do so, although once out of the
yard it may stop jumping at any desired spot. After a counter
has left its yard it may not rest in the yard again, although it
may hop across it.

The Halma problem devised by Matsuda for the Japanese
chessboard, which has nine cells on a side, begins with nine
counters in a square array at the board's lower left corner.
How few moves of the Halma type, Matsuda asked himself, are
needed to transfer the nine counters to the same formation at

the upper right corner? He found a solution in 17 moves, but this was reduced to 16 moves [*see Figure* 67] by H. Ajisawa and T. Maruyama. The 16-move solution is believed to be minimum.

Figure 67

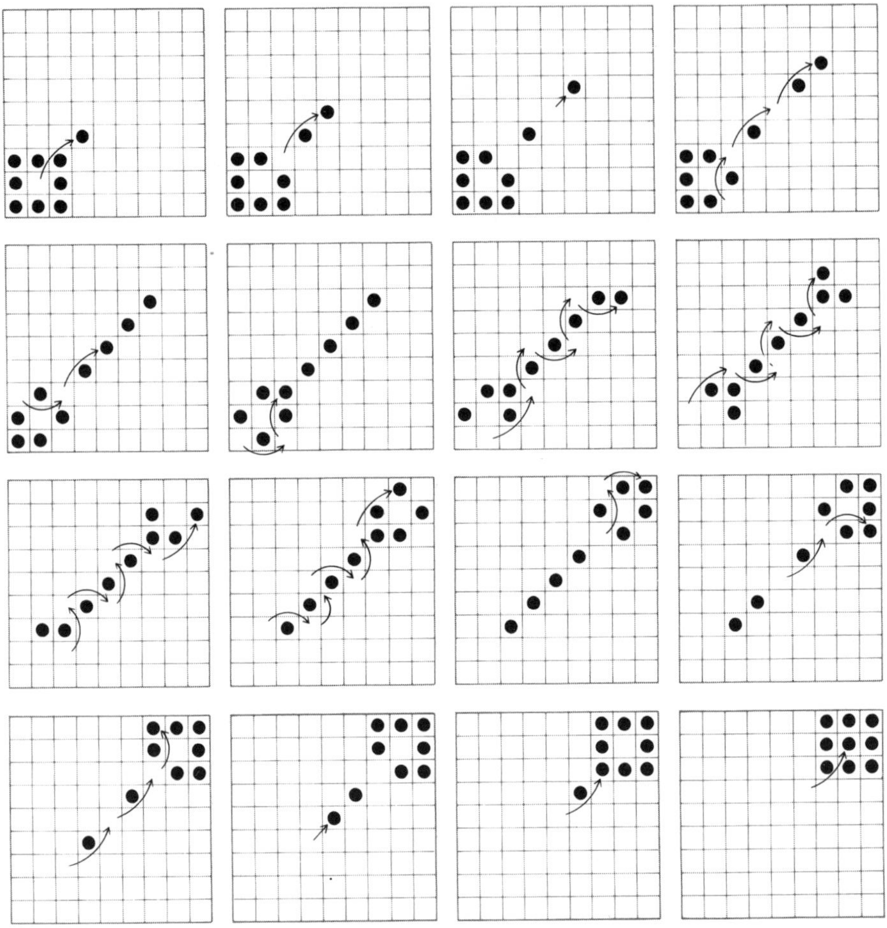

Solution to Matsuda's problem on the Japanese chessboard

When I saw this elegant solution, I at once began tackling the same problem on the Western chessboard with eight cells on a side and on smaller square boards with seven and six cells on a side. In each case, a square of 3×3 counters is diagonally shifted to the opposite corner. Using the technique of first establishing a diagonal ladder—a basic strategy, by the way, of all games of the Halma type—the best I could achieve was 15 for the chessboard, 13 for the order-7 and 12 for the order-6. I have been unable to prove that any of these are minimum solutions. It is not hard to show that at least 12 moves are necessary for the order-8 square, 10 for the order-7 and 11 for the order-6.

Next I experimented with a similar transfer of the nine-counter square, on the same three boards, to the lower right corner instead of the corner diagonally opposite. The order-6 board has many solutions in nine moves, one of which is shown in Figure 68. Nine is obviously minimal because each counter must move once. (It is necessary that at least one counter hop to and from the fourth row on its way to the other yard, consequently nine-move solutions cannot be achieved on a three-by-six board.) On the order-7 board 10 moves will do it. This too is readily seen to be minimal since the first piece to move must move at least once again to reach the adjacent yard.

Figure 68

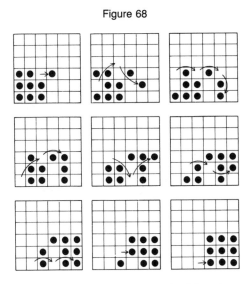

Orthogonal transfer on an order-6 board

Thirteen moves will solve the problem on the order-8 board. That 12 are necessary is evident from a simple parity check. The six counters in column 1 and column 3 can hop only to column 7, therefore three of the six must each make at least one step move. I tried vainly for weeks to find a 12-move solution until Donald E. Knuth, a mathematician at Stanford University, came to my rescue by devising a proof of impossibility in 12 moves. It is too involved to give here, but it is based on the necessity for one of the original four corner counters to step to a different color, the fact that the reverse of a solution is another solution and other considerations. Readers may enjoy searching for minimum solutions to the six transfer problems.

The second family of puzzles suggested by Matsuda's problem is based on removing every jumped counter from the board. The goal is to remove all counters but one, the last counter reaching a specified cell, and do it in a minimum number of Halma moves. Such problems are similar to those of the classic peg-solitaire game discussed in an earlier column (reprinted in my book *Unexpected Hanging and Other Mathematical Diversions*) except that the greater freedom of movement allows for much shorter solutions, and proofs of minimum solutions are usually much more difficult.

Consider, for example, the puzzle on a five-square board that was first issued in 1908 by Sam Loyd [*see Figure* 69, *No. 1*]. He labeled each counter with the name of a hopeful in that year's presidential election. The idea was to eliminate eight men, leaving one's favorite on the center cell. Loyd allowed Halma moves but did not count a chain of jumps as being one move. Eight jumps are clearly minimal and there are many such solutions for each counter. Henry Ernest Dudeney, in his *Amusements in Mathematics* (Problem 229), improved the puzzle by disallowing step moves, counting jump chains as single moves and allowing any counter to end at the center. He gave a four-move solution that is surely minimal, although I know of no proof. Counter 5 jumps 8, 9, 3, 1; counter 7 jumps 4; 6 jumps 2 and 7; then 5 returns to its original cell by leaping 6.

Figure 69

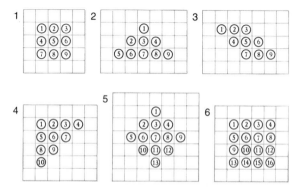

Six Halma solitaire problems

Let us combine the rules of the two rival puzzlists by allowing both steps and hops, as in Halma, and counting a chain of hops as one move. Each hopped counter is of course removed. Can the reader find one of the many three-move solutions that leave the last counter on the center cell? The solution is an elegant one that begins with two step moves and ends with an eight-jump chain.

Similar problems are shown in the same illustration, numbered 2 through 6. The second is to be solved in three Halma moves, the surviving counter on the cell initially occupied by the counter at the top of the triangle. The third problem is to be solved in three moves, last counter on the board's center cell. The fourth problem is to be solved in a minimum number of moves, last counter on the cell initially occupied by counter 6, the triangle's center. The fifth problem is to be solved in three moves, surviving counter at the board's center. The last problem, the most difficult of the six, calls for three moves that end with the lone counter on one of the board's four center cells.

The field of Halma puzzles is so unexplored that it is a challenge to devise and solve new puzzles, then see if one can prove by simple arguments that the solution actually is minimal. I have not the slightest notion, for example, how few moves are required on an order-7 board with 25 counters in a square array in the center to leave the last counter on the center cell. I have avoided trying this problem for fear of accomplishing no other work for the next month or so.

ANSWERS

The six Halma problems can be solved as follows. None of the solutions is unique:

1. Counter 6 steps diagonally up and right, 8 (or 4) steps diagonally down and left, 5 jumps all counters to end at the center.

It is possible in three all-jump moves (no steps) to end on either a corner cell or a side cell of the original pattern, but when steps are not allowed, four moves are necessary (they were given earlier) to reach the center. Two moves suffice to remove eight counters but the survivor will be outside the original pattern.

2. Counter 4 steps up, 3 jumps 8, 9, 4, 1, 2, 5, 6, then 7 jumps counter 3. A three-move solution that puts the last counter on the board's center cell is: 4 steps up, 6 steps down, and 3 jumps the remaining eight counters. A neat symmetrical solution: 1 steps up and 7 jumps 3 (or 7 steps down and 1 jumps 3), then 1 jumps the rest.

3. Counter 6 steps up, 8 (or 4) steps down, 5 jumps all counters to rest on the center cell. This pattern and its solution are equivalent to the first problem, with each diagonal move changed to vertical and each vertical move to diagonal, all horizontal moves remaining the same. There are similarly equivalent patterns and solutions on the checkerboard and the Chinese checkers board.

4. Counter 6 can jump all counters in one move, returning to its original cell at the center. The problem is equivalent to a 10-counter equilateral triangle on the Chinese checkers board.

5. Counter 11 hops diagonally up and right (eliminating counter 8), 6 jumps 10 counters and returns to its former cell, then 5 removes 6 as it leaps to the center.

6. Counter 8 steps diagonally up and right, 14 jumps 9, 1, 3, 11 and returns to its former spot, then 8 jumps 11 counters to end on the cell originally occupied by 11.

Another problem, a three-by-four rectangle on a five-by-six field, can be solved in three all-jump moves, the final counter resting on any of the 12 cells of the original pattern. In two moves the board can be cleared but the last counter will be outside the original pattern.

ADDENDUM

Five readers (Katsumi Takemura, Seiichi Fusamura, Mitsunobu Matsuyama, James Stuart, and Y. Dvir) lowered to 12 the moves required to transfer the 3×3 square of counters diagonally from corner to corner on the order-7 board.

The three-move solution given for the Halma solitaire problem involving the order-4 array on an order-6 board apparently not only is fundamentally unique for ending on one of the central cells but also seems to be the only three-move solution that eliminates all but one counter when this last counter can end anywhere on the board. When the order-4 square is at the center of a standard eight-by-eight chessboard, a pretty four-move solution puts the last counter on a corner of the board. And I found a four-move solution that leaves four counters at the corners of the order-4 square when it is centered on the order-6 field.

Here are some more of my results: The order-5 array (on the order-7 board) has four-move solutions that end on any cell originally occupied; the order-6 formation (on the order-8 board) has six-move solutions to any cell formerly occupied, and a five-move solution to the board's corner, and there are two-move solutions for the order-3 array that place the last counter on any cell of the order-5 board's border.

John W. Harris was the only reader to send results for the order-7 array on the order-9 Japanese chessboard. He found a solution, to the center cell, in seven moves.

If a square array of nine counters are placed at a corner of a 4×6 board, it is a pleasant task to shift them to the diagonally opposite corner in ten moves. The small size of the board makes it an attractive puzzle to market. I offer it free to any firm that cares to manufacture it, either with marbles, counters, or pegs in holes. I found a solution in ten moves and proved it to be minimal.

BIBLIOGRAPHY

The Handbook of Reversi, also Fanorona, Invasion, Halma. London: F. H. Ayres, 1889.

The Book of Table Games. Edited by "Professor Hoffmann" (Angelo Lewis). London: George Routledge and Sons, 1894.

12

ADVERTISING PREMIUMS

Inexpensive advertising premiums are popular in all countries where businesses compete for consumer attention, and frequently such premiums are based on mathematical puzzles. Many premiums of this kind have been discussed in columns that are reprinted in my earlier book collections, and one involving a "map fold" will be found in this book in the chapter on paper folding. Now I shall consider some classic puzzle premiums that I have not previously discussed.

One of the oldest and best is the *T*-puzzle shown in Figure 70. The reader is urged to trace or photocopy the four pieces, paste them on cardboard, cut them out and try to fit them together to make a capital *T*. I know of no polygon-dissection puzzle with as few pieces that is so intractable. The number of giveaway premiums based on this puzzle, particularly in the early decades of the century, runs into the hundreds.

Figure 70

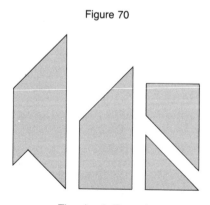

The classic T-puzzle

Less well known, although equally ancient and charming, is the square puzzle shown in Figure 71. It is best shown to a friend by first giving him only the four nonsquare pieces and asking him to make a square. After he succeeds, hand him the square piece and see how much longer it takes him to make a square that uses all *five* pieces.

Figure 71

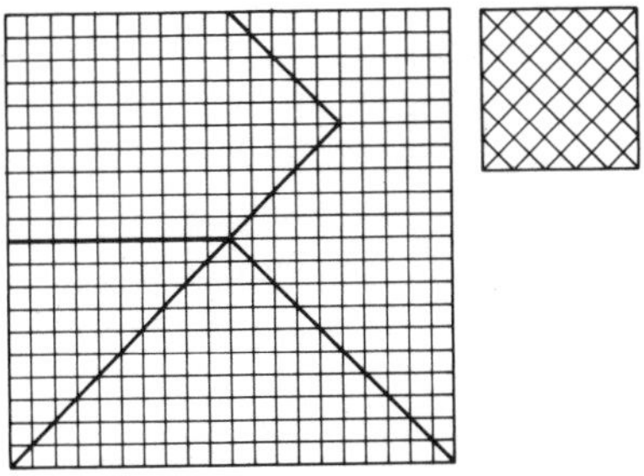

The Pythagorean-square puzzle

I have not seen this as a die-cut premium in recent years, but at least two plastic versions are currently on sale in the United States. Milton Bradley's One Way was designed by Henry Adams, and another version, designed by Frank Armbruster, is called Madagascar Madness. Armbruster's instructions point out how the puzzle illustrates the Pythagorean theorem. If the big and little squares shown in the illustration are on the sides of a right triangle, the square formed by all five pieces will, of course, be the square on the hypotenuse.

In this country the most prolific creator of mathematical premiums unquestionably was Sam Lloyd (1841–1911), the famous Philadelphia-born puzzlist and chess-problem inventor. In his cluttered, musty office in a decaying Manhattan building occupied by *The Evening Globe,* Loyd concocted hundreds of puzzles of fantastic variety and ingenuity. As described in a 1911 magazine article, his small office "would be dark even if the one window were washed, a cataclysm of which there seems no immediate prospect. There are two desks, a typewriter and a printing-press in it, and countless shelves loaded with papers, pictures, magazines, stereotype plates and one thousand other things which have spilled out upon the floor and risen like strange, dirty snowdrifts breast high in the corners. Sam Loyd says he does all his business on a cash basis and keeps no books. The reason probably is that he couldn't find the books. That would be too much of a puzzle even for him."

Loyd's first big success with a premium came with his invention, at the age of 17, of the Trick Donkeys. The task is to arrange three cardboard rectangles so that two riders are astride two donkeys. The puzzle is still widely used as a giveaway item throughout the world. Loyd's original version, which P. T. Barnum distributed by the millions to publicize his circus, is reproduced in the chapter on Loyd in *The Scientific American Book of Mathematical Puzzles & Diversions.* Modernized versions can be found in the article "Problem-solving" in *Scientific American* for April 1963; on page 124 of *The Mind* (a Life Science Library book), and in an American Can Company advertisement in *Time* for March 22, 1968. Loyd once related in an interview that Barnum used to make periodic treks to his office saying, "Hang it all, Sam, show me how to do my puzzle. I've forgotten again."

Another of Loyd's early premium hits, even more widely used today than then, is nothing more than a pencil with a short loop of string on its eraser end. Loyd designed the trick for agents of the New York Life Insurance Company, who would attach the pencil to prospective customers' coats with the promise to remove it if a sale was consummated. The loop is placed around a lapel buttonhole, then the cloth is pulled forward through the loop until the pencil goes back far enough for its point to enter the buttonhole from behind. When the pencil is pulled foward through the hole, it is fastened in such a way that it seems impossible to remove the pencil without cutting the string.

Loyd produced numerous geometric puzzles, but none with a more unexpected solution than his Pony Puzzle, shown in Figure 72 exactly as he himself originally drew it. The problem is to rearrange the six pieces to make the best possible picture of a trotting horse. In his *Cyclopedia of Puzzles* Loyd claimed that over one billion copies of the Pony Puzzle had been sold.

The most spectacular of all Loyd premiums, by all odds, was his mind-bending "Get off the Earth" paradox. He patented the device in 1896 and first sold it as a premium to advertise Bergen Beach, a resort that had just opened in New Jersey. Copies of the original are now rare collector's items. The art, based on Loyd's sketches, was done by Anthony Fiala, then a cartoonist on *The Brooklyn Daily Eagle.* (Later Fiala was commander of the Ziegler Polar Expedition of 1903–1905 and wrote a book about it, *Fighting the Polar Ice.*) The puzzle consisted of a cardboard disk fastened by a central rivet to a card-

Figure 72

Sam Loyd's Pony Puzzle

board rectangle. A tab attached to the disk projected through a curved slot in the backing so that by moving the tab up or down the disk could be rotated to two positions [*see Figures* 73, 74]. In one of the positions you can count 13 Chinese warriors. When the disk is turned to the other position, there are only 12 warriors. Which man vanishes, the premium asked, and where does he go?

For more than a year Loyd filled his weekly puzzle column in *The Brooklyn Daily Eagle* with letters from readers attempting to explain this astonishing phenomenon. In Loyd's own lengthy, mock-serious explanation (January 3, 1897, page 22) he called attention to a curious feature easily overlooked by a person unless he has tried the difficult task of drawing human figures properly around the rim of a disk. "The grotesqueness of the figures and a necessary legerdemain feat of changing a right leg for a left one between the fourth and fifth men does the trick. If it were not for that particular acrobatic feat, all of the men on the left side would come down head end first. Some pirates, who brought out the puzzle in different parts of Europe, with different figures, found it absolutely necessary to retain that flop over of the legs."

At that time Americans were aroused over the "yellow peril," a fact that explains the premium's unpleasant racist connotations. As if not to be partial to either China or Japan (the two nations had been at war in 1894), Loyd provided the Metropolitan Life Insurance Company in 1897 with a more elaborate Japanese version of his paradox. Nine Japanese men alternate around the circle with eight lanterns. When the disk is turned, one man vanishes and a ninth lantern appears, giving the

Figure 73

Loyd's greatest puzzle starts with 13 Chinese warriors

Figure 74

Now there are only 12 warriors. Which one disappears?

impression that a man has turned into a lantern. The premium announced a contest with 20 prizes, from $5 to $100, for the best explanation. Although the names of the winners were printed, none of the prize-winning letters were published. Perhaps the reason is to be found in a typical "explanation" that was quoted: "When the handle is down I find nine Japanese, but when the handle is up there are only eight, as one has disappeared." In 1909 Loyd issued a third version of the paradox called Teddy (Roosevelt) and the Lions, in which an African native seems to turn into a lion. It too is reproduced in the chapter on Loyd in *The Scientific American Book of Mathematical Puzzles & Diversions.*

The basic principle behind Loyd's three versions was not original with him. He simply took earlier linear forms of the paradox and bent them into circular shape. I have seen in a private collection of advertising cards an 1880 premium, copyrighted by Wemple and Company of New York, called "The Magical Eggs." A rectangular card is cut into four smaller rectangles. Different arrangements of the pieces produce eight, nine or 10 eggs. Scores of variations on this paradox have since been used in the United States and abroad. The latest and funniest version, in three pieces, is "The Vanishing Leprechaun," skillfully drawn by Pat Patterson, a Toronto graphic designer, and issued in Canada by William Elliott, a producer of puzzles and magic tricks. The paradox is repro-

Figure 75

Which leprechaun vanishes?

duced in Figure 75. An eight-by-19-inch two-color print on paperboard can be obtained from W. A. Elliott Company, 212 Adelaide Street West, Toronto 1, Canada.

From hundreds of other mathematical premiums I select as a final specimen a card that advertises a brand of Scoth whisky [*see Figure* 76]. This seemingly trivial addition problem trips most people whether they have had a drink or not. To obtain the correct sum the use of a pocket or desk calculator is advised.

Figure 76

Can you add this column of figures? Place your hand over all but the top number and move it down the column, revealing one number at a time. Add all the numbers, as you go along. When you get the total, turn over for correct answer.

1000

40

1000

30

1000

20

1000

10

An advertising giveaway card

ANSWERS

Solutions to the two dissection puzzles, and Sam Loyd's pony puzzle, are shown in Figures 77, 78, and 79. I leave it to the reader to decide which leprechaun vanishes and where the little fellow hides.

Figure 77

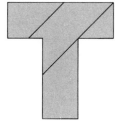

Solution to the T-puzzle

Figure 78

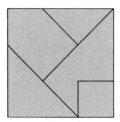

Solution to the square puzzle

Figure 79

Solution to Sam Loyd's Pony Puzzle

ADDENDUM

Manuel R. Pablo, of the Naval Research Laboratory, Washington, D.C., surprised me by finding another solution to the old *T* puzzle. By turning one piece over he produced the fat *T* shown in Figure 80. Other readers, keeping the five-sided piece in its standard orientation, produced *T*s with arms of different lengths.

Figure 80

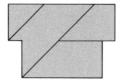

Pablo's solution to the T-puzzle

Note that if one piece is turned over, the four pieces fit neatly together to form the isosceles trapezoid shown in Figure 81. The *T* puzzle, packed in this trapezoidal form, was on sale in 1975 as the "Teezer" puzzle, made by Hoi Polloi, New York City.

Figure 81

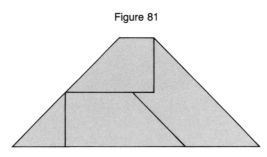

The "Teezer" puzzle

The *T* puzzle has been made with the *T* in many different shapes, but the puzzle is difficult only if the five-sided piece has the same width as the others. The mind has a strong tendency to assume that this piece must go either vertically or horizontally, an assumption that of course makes the solution impossible.

David Frost was so intrigued by the leprechaun paradox that he arranged for Pat Patterson to provide an enlargement that he could display on the TV talk show he was then hosting. After demonstrating the paradox, Frost asked if anyone in the audience could explain it. Nobody could. Finally a lady stood up to say that her husband understood how it worked. Frost turned to the husband. His explanation was identical with the one I quoted for Loyd's vanishing Chinese warrior. When the rectangles are arranged one way, the man said, there are 15 leprechauns. But when you arrange them the other way, there are only 14.

In showing the paradox to friends, an amusing bit of business is to ask which leprechaun vanishes. If they pick one, put a penny on the upper half and another penny on the lower half. After shifting the pieces, the pennies of course mark portions of leprechauns that are still there. Let them try again. The number of pennies on the figures increase, but without casting much light on the mystery.

Many readers wrote to say that if the lower half of the picture is cut in two parts by a vertical cut between the ninth and tenth leprechauns, you can arrange the four pieces to make 13 leprechauns. Other permutations produced by other cuts will give 16 and 17 figures, though they get distorted as they increase. Of course you can produce similar changes by rotating Loyd's disk.

Dozens of imitations and variations of the leprechauns have been printed since the item was first marketed, some of them pornographic. You will find an early discussion of how they all work, with many examples, in my *Mathematics, Magic and Mystery,* and in Mel Stover's cover article, "The Disappearing Man and Other Vanishing Paradoxes," listed in the bibliography. Stover owns the largest collection of such things, mine running a close second.

BIBLIOGRAPHY

On Sam Loyd:

"The Prince of Puzzle Makers." George Grantham Bain. *The Strand Magazine,* Vol. 34, December 1907, pages 771–777.

"My Fifty Years in Puzzleland: Sam Loyd and His Ten Thousand Brain-Teasers." Walter Prichard Eaton. *The Delineator,* April 1911, page 274.

The Mathematical Puzzles of Sam Loyd. Edited by Martin Gardner. Vol. 1, Dover, 1959; Vol. 2, Dover, 1960.

On vanishing figures:

Mathematics, Magic and Mystery. Martin Gardner. Dover, 1956.

"The Disappearing Man and Other Vanishing Paradoxes." Mel Stover. *Games,* November–December 1980, pages 14–18.

13

SALMON ON AUSTIN'S DOG

In Chapter 8 one of the short problems, posed by A. K. Austin of the University of Sheffield, England, aroused considerable controversy among readers. Indeed, the problem proved to be an amusing new variant of Zeno's famous paradox of Achilles and the Tortoise, and one that, so far as I know, had never been formulated before. Here is how I phrased the problem and its answer:

> "A boy, a girl and a dog are at the same spot on a straight road. The boy and the girl walk forward—the boy at four miles per hour, the girl at three miles per hour. As they proceed, the dog trots back and forth between them at 10 miles per hour. Assume that each reversal of its direction is instantaneous. An hour later, where is the dog and which way is it facing?"
>
> Answer: "The dog can be at any point between the boy and the girl, facing either way. Proof: At the end of one hour, place the dog anywhere between the boy and the girl, facing in either direction. Time-reverse all motions and the three will return at the same instant to the starting point."

Even before this answer appeared I began receiving letters from readers protesting that the problem is meaningless because its initial conditions are logically contradictory. No matter how small we make the starting interval, many wrote, the dog will have to make an infinity of reversals that would drive it crazy. Others contended that the three "points" (as in all such problems, the boy, the girl and the dog symbolize ideal points) could never get started because the "instant" they did so the dog would either leap ahead of both boy and girl or run the opposite way, thereby ceasing to be between the boy and girl.

As Wesley C. Salmon, a noted philosopher of science, immediately recognized, Austin's paradox has innumerable other forms, one of the simplest of which is a time reversal of the familiar puzzle about two locomotives and a bird. The locomotives, starting at A and B, 30 miles apart, move toward each other on the same track at, say, 15 miles per hour until they collide at C. A bird, starting at A, flies back and forth at 60 miles per hour between the locomotives until they collide. How long is the bird's path? There is no need to sum an infinite series. Since the bird flies for one hour, the path must be 60 miles. If we time-reverse the event, specifying that the bird end at A, a unique zigzag path is defined that the bird can travel in either direction.

Suppose, however, we do not state where the bird must be after the locomotives have moved backward to points A and B. Without this information a unique path for the bird cannot be defined. Because the bird can now take an infinity of possible paths, the most we can say is that the backward-flying bird must end somewhere between A and B.

But is it really permissible to say this? No, many mathematicians contend, because a singularity arises in the time-reversed version that creates contradictory initial conditions. "There is no general justification in analysis," one mathematician put it, "for inverting the limit operator." When the locomotives move toward each other, it is only the bird's position that converges. "The velocity vector diverges, so that there is the same difficulty (as in Austin's problem) in finding a unique inverse to the limit process. The accepted rules of differential calculus have evolved because if followed properly they avoid contradictions."

It is helpful to plot a space-time graph of the bird's path from A to C' [see Figure 82]. Of course, we cannot finish drawing the bird's path to C' because the zigzags are infinite, but we certainly can assume that the ideal line exists. Surely if this line can go down from A to C', there is no logical objection to saying that it can go up from C' to A. If the final destination of the bird is not specified, an uncountable infinity of such graphs can start at C' and end anywhere on the track between A and B. It is true that calculus cannot solve Austin's similar problem if "solve" means to pinpoint the dog's final position, but Austin's "solution" is precisely one that shows this to be impossible. Since the dog is not told how to start, it can start in any way it pleases provided it always stays between the boy and the girl. Consequently its path can end anywhere between boy and girl.

Salmon has commented on Austin's problem as follows:

Figure 82

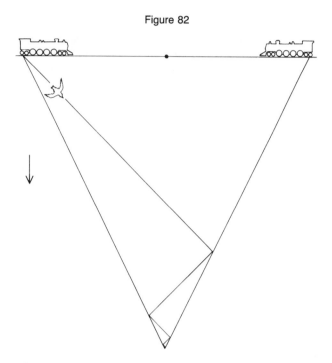

Space-time graph of bird's path between moving locomotives

"Almost everyone has heard the old chestnut about the bird that flies back and forth between two approaching locomotives ... [as given above]. Or, to achieve historical perspective, suppose Achilles is pursuing the tortoise and a Trojan fly buzzes back and forth between them. Given a set of velocities and distances, and our latter-day assurance that Achilles will overtake the tortoise at a determinate time and place (see my book *Zeno's Paradoxes*), we can easily figure out how far the fly will travel. Up to this point we have no new Zenonian paradoxes. . . . We see that Austin's problem is just the time-reversal of the bird-and-train problem.

"In order to retain historical perspective, let us go back to Achilles and the tortoise. In spite of the initial handicap traditionally imposed on Achilles, he catches the tortoise, and to redress the grievance he has long held against Zeno he keeps on running, steadily increasing his lead over the fortunate tortoise. [I consider the tortoise fortunate in this version of the tale, at least in comparison with Lewis Carroll's account "What the Tortoise Said to Achilles," in which Achilles stops and seats

himself on the back of the tortoise, much to the tortoise's dis-
comfort.] Now consider the Trojan fly, which attempts to continue flying back and forth between the two runners even after
the faster overtakes the slower. When Achilles and the tortoise
are just even, the fly finds itself precisely in the position of
Austin's dog.

"For the sake of definiteness, say that the tortoise travels at
one mile per hour, Achilles at five miles per hour (he has been
running since 500 B.C., so that he is not as fleet as he once was)
and the fly at 10 miles per hour. They all arrive at the common
meeting point without difficulty. How can they go on? If the
three start simultaneously from the common point, the fly immediately either advances ahead of both or moves behind both,
each of which violates the condition that the fly be always in
the interval between the two (end points included). It would
seem we could argue that in any time interval $\varepsilon > 0$, however
small, the tortoise travels a distance of 1ε, Achilles runs a distance of 5ε and the fly goes 10ε. Hence in an arbitrarily small
time after the meeting the fly leaves the interval between the
tortoise and Achilles. Even if we have shown how Achilles can
perform the 'supertask' of catching the tortoise, and how the
tortoise can perform the 'supertask' of initiating its motion, it
appears that the fly now faces the new 'supertask' of continuing
to fly back and forth between Achilles and the tortoise after the
tortoise has been overtaken. In other words, the fly now faces
the supertask of *not* passing Achilles!

"The apparent difficulty seems to me analogous to the problem pointed out by Zeno in his regressive dichotomy paradox.
There is no doubt that the fly will outdistance both Achilles
and the tortoise *if it flies steadily in one direction without turning
around,* even in the arbitrarily small period of duration ε. This
fact does not render the fly's motion impossible, however, since
no matter how small a time interval we choose the fly has already reversed its direction during that interval (infinitely
many times, so that it is really quite dizzy). This simply means
that there is no *initial* nonzero interval during which it flies
straight without reversing its direction; thus it does not follow
that the fly immediately leaves the interval between the tortoise
and Achilles. In fact, we can see precisely how the fly's rapid
reversals enable it to stay between Achilles and the tortoise
after the meeting by examining the time reversal of this motion
in the fly's approach to the point of meeting from the earlier
side. The fact that the fly does not traverse an *initial* nonzero
straight path is analogous to the fact that the tortoise, in leav-

ing its starting point, does not traverse any initial nonzero segment of its path. The lack of a suitable initial segment is not a serious obstacle to either of them.

"The recent literature on Zeno's paradoxes has contained a good deal of discussion of 'infinity machines.' These are idealized devices that purportedly perform an infinite sequence of tasks; they have been introduced into the discussion because of difficulties they seem to encounter in completing the infinite sequence of tasks (a 'supertask'). The resolution of the problems surrounding the infinity machines is strongly analogous to the resolution of the progressive form of Zeno's dichotomy paradox. The motion of the Trojan fly up to and including the moment Achilles overtakes the tortoise involves exactly the same considerations. I am not aware that anyone has explicitly introduced the kind of infinity machine that would be analogous to the regressive form of Zeno's dichotomy paradox, a machine whose difficulty lies in getting started with its series of tasks, in contrast with the usual infinity machine whose difficulty lies in finishing its series of tasks. As it turns out, our Trojan fly, in its motion from the point of meeting of Achilles and the tortoise through the subsequent part of the run in which Achilles is ahead of the tortoise, constitutes just such an infinity machine (as does Austin's dog)—a regressive infinity machine, we might say. Just as the treatment of the standard infinity machine closely parallels the resolution of the progressive dichotomy paradox, so does the treatment of the Trojan fly in the latter part of its flight closely parallel the resolution of the regressive dichotomy paradox.

"One further problem about the motion of the fly deserves explicit attention, namely what is the state of motion of the fly at the precise instant of meeting? The fly's position is well determined; it coincides with the position of Achilles and the tortoise. The mathematical function that describes the fly's position is a continuous function of time that passes through the meeting point at the appropriate instant. The fly's velocity function, on the other hand, is discontinuous. Its value is $+10$ when the fly is moving forward, -10 when it is moving backward and (we might as well say) zero when the fly meets either Achilles or the tortoise (or both). Thus we can appropriately assign the value zero to the fly's velocity at the instant when all three meet. Obviously the velocity function has infinitely many discontinuities on each side in the neighborhood of the point of common meeting. Each *finite* discontinuity in the velocity function corresponds to an *infinite* discontinuity in the acceleration, since it requires an infinite acceleration for the fly to change velocity instantaneously from $+10$ to -10 and vice versa. Moreover, as Austin's problem and its solution show, the state of motion of the fly (or dog) at the point of meeting does not uniquely determine how the motion is to continue beyond that point. In other words, although we have shown how (in some sense of 'possible') it is possible for the fly to continue its motion through the meeting point and beyond, the motion beyond the meeting point can be executed in infinitely many distinct ways, all of which are consistent with the conditions imposed by the problem. To say that there are alternative ways of performing a task does not, however, prove that the task is impossible to execute.

"In the customary formulations Zeno's Achilles and dichotomy paradoxes involve a finite number of discontinuities of the type just mentioned: Achilles and the tortoise are assumed to accelerate instantaneously at their starting points to their respective average velocities, and to decelerate instantaneously to zero at the finish. Similarly, most of the 'infinity machines' (for example Black's transferring machines and the Thomson lamp) involve infinitely many such discontinuities clustering around some moment of termination (see *Zeno's Paradoxes*, pages 204–244). Using a mathematical function supplied by Richard Friedberg, Adolf Grünbaum has shown how such motions can be modified so as to eliminate all the discontinuities and still achieve the desired total outcome. It seems reasonable to conjecture that a similar approach could be applied to the problem of the Trojan fly (or Austin's boy-girl-dog) in order to achieve a totally unobjectionable description of the motion."

ADDENDUM

I had expected Professor Salmon's analysis of Austin's paradox to produce many letters of disagreement, but evidently Salmon argued his case skillfully because I received not a single one. Of course the debate is largely verbal, a question of what sort of language to use in making the problem and its solution precise.

Many other problems are analogous to Austin's dog in the sense that there is a precise answer in forward time, but hopeless ambiguity when the event is time reversed. Consider for instance a point starting at the earth's equator and moving due north with uniform speed along a loxodrome. It will circle the north pole a countable infinity of times, reaching the pole at a precise instant. But time-reverse the event and the point can cross the equator at any spot. Because there is no "last" revolution around the pole, there is no precise beginning of the time-reversed event that will determine a unique spiral path.

Mathematics Magazine, which originally published Austin's paradox as problem $Q503$ (January 1971), returned to the paradox in its September issue by publishing comments by four mathematicians, all of whom considered the problem self-contradictory. The magazine did not publish Salmon's reply to one comment. I reproduce it below:

In the September–October number, Lyle E. Pursell comments on Quickie $Q503$ (Austin's boy-girl-dog problem) as follows:

> The author's solution to the problem looks like a proposal to sum an infinite series by starting at the "last" term! Since, if the latter three reverse their motions as the author suggests in his solution, then the dog must reverse his direction infinitely many times before the boy and the girl get back to the starting point.

While no original texts have survived to the press date, it seems plausible to suppose that Zeno of Elea (circa 500 B.C.) might have made a similar comment about Achilles:

> Since Achilles must run half of the racecourse before he can run the whole, and he must run a quarter before he can complete the half, etc., it is evident that Achilles must run infinitely many distances before he can have reached any point, however near, beyond his starting point. To say that Achilles has run any finite (i.e., nonzero) distance looks like a proposal to sum an infinite series starting at the "last" term!

Although Austin's dog must reverse his direction between segments whereas Zeno's Achilles keeps going in the same direction, does this difference really have any bearing upon the absurdity involved in the "proposal to sum an infinite series by starting at the 'last' term!"? It appears that Austin's dog exhumes Zeno's old regressive dichotomy paradox. If Achilles can run a racecourse, why cannot Austin's dog do what is required of him?

BIBLIOGRAPHY

Zeno's Paradoxes. Wesley C. Salmon (ed.) Bobbs-Merrill, 1970.

Modern Science and Zeno's Paradoxes. Adolf Grünbaum. Wesleyan University Press, 1967.

"Comment on Q503." M. S. Klamkin, Leon Bankoff, Charles W. Trigg, and Lyle E. Pursell. *Mathematics Magazine,* Vol. 44, September 1971, pages 238–239.

Space Time & Motion. Wesley C. Salmon. Dickenson, 1975. Secon revised edition, University of Minnesota Press, 1980, pages 48–52 (both editions).

"Mary, Tom, and Fido." Martin Gardner in *Aha! Gotcha.* W. H. Freeman, 1982, pages 148–149.

14

NIM AND HACKENBUSH

"The good humour is to steal. . . ."

—WILLIAM SHAKESPEARE, Corporal Nym in
The Merry Wives of Windsor

In recent decades a great deal of significant theoretical work
has been done on a type of two-person game that so far has no
agreed-on name. Sometimes these games are called "nim-like
games," "take-away games" or "disjunctive games." All begin
with a finite set of elements that can be almost anything:
counters, pebbles, empty cells of a board, lines on a graph, and
so on. Players alternately remove a positive number of these
elements in accordance with the game's rules. Since the elements diminish in number with each move, the game must
eventually end. None of the moves is dictated by chance; there
is "complete information" in that each player knows what his
opponent does. Usually the last player to move wins.

The game must also be "impartial." This means that permissible moves depend solely on the pattern of elements prior to
the move and not on who plays or on what the preceding
moves were. A game in which each player has his own subset
of the elements is not impartial. Chess, for example, is partial
because a player is not allowed to move an opponent's piece. It
follows from the above conditions that every pattern of elements is a certain win for either the first or the second player
if the game is played rationally. A pattern is called "safe" (or
some equivalent term) if the person who plays next is the loser
and "unsafe" if the person who plays next is the winner. Every
unsafe pattern can be made safe by at least one move, and

every safe pattern becomes unsafe through *any* move. Otherwise it is easy to prove the contradictory result that both players could force a win. The winner's strategy is playing so that every unsafe position left by the loser becomes a safe one.

The best-known example of such a game is nim. The word was coined by the Harvard mathematician Charles L. Bouton when he published the first analysis of the game in 1901. He did not explain why he chose the name, so we can only guess at its origin. Did he have in mind the German *nimm* (the imperative of *nehmen*, "to take") or the archaic English "nim" ("take"), which became a slang word for "steal"? A letter to *The New Scientist* pointed out that John Gay's *Beggar's Opera* of 1727 speaks of a snuffbox "nimm'd by Filch," and that Shakespeare probably had "nim" in mind when he named one of Falstaff's thieving attendants Corporal Nym. Others have noticed that NIM becomes WIN when it is inverted.

Nim begins with any number of piles (or rows) of objects with an arbitrary number in each pile. A move consists in taking away as many objects as one wishes, but only from one pile. At least one object must be taken, and it is permissible to take the entire pile. The player who takes the last object wins. Bouton's method of determining whether a nim position is safe or unsafe is to express the pile numbers in binary notation, then add them without carrying. If and only if each column adds to an even number (zero is even) is the pattern safe. An equivalent but much easier way to identify the pattern (with practice one can do it in one's head) is to express each pile number as a sum of distinct powers of 2, eliminate all pairs of like powers and add the powers that remain. The final sum is the nim sum of the pattern. In current parlance this is called the "Grundy number" or "Sprague-Grundy" number of the pattern, after Roland Sprague and P. M. Grundy, who independently worked out a general theory of take-away games based on assigning (by techniques that vary with different games) single numbers to each state of the game.

For example, assume that a game of nim begins with three piles of three, five and seven counters.

$$3 = 2 + 1$$
$$5 = 4 + 1$$
$$7 = 4 + 2 + 1$$

Pairs of 4's, 2's and 1's are crossed out as shown. The sum of what remains is 1. This is the nim sum of the pattern. If and only if the nim sum is zero is the pattern safe, otherwise it is unsafe (as it is here). If you play an unsafe pattern, you win by changing it to safe. Here removing one counter from any pile will lower the nim sum to zero. In three-pile nim, with no pile exceeding seven counters, the safe nim patterns are 0–*n*–*n*, where *n* in the first triplet is any digit from 1 through 7, and 1–2–3, 1–4–5, 1–6–7, 2–4–6, 2–5–7, 3–4–7, 3–5–6. If your opponent plays next, he is sure to leave a pattern with a nonzero nim sum that you can lower to zero again, thereby maintaining your winning strategy.

Like all games of this type, nim has its *misère* form, in which the player who takes the last piece is the loser. In many take-away games the strategy of *misère* play is enormously complicated, but in nim only a trivial modification is required at the end of the play. The winner need only play a normal strategy until it is possible to leave an odd number of single-counter piles. This forces his opponent to take the last counter.

Many take-away games seem to demand a strategy different from that of nim but actually do not. Suppose the rules of nim allow a player (if he wishes) to take from a pile, then divide the remaining counters of that pile into two separate piles. (If the counters are in rows, this is the same as taking contiguous counters from inside a row and regarding those that remain as being two distinct rows.) One might expect this maneuver to complicate the strategy, but it has no effect whatever. To win, compute the nim sum of a position in the usual way and, if it is unsafe, play a standard move to make it safe. For example, in the 3–5–7 game suppose your first move is taking a counter from the three-pile, leaving the safe 2–5–7. Your opponent removes two counters from the seven-pile and splits the remaining five counters into a two-pile and a three-pile. The pattern is now 2–5–2–3. Its nim sum is six, which you make safe by taking two from the five-pile.

A pleasant counter-moving game on a chessboard is shown in Figure 83. No fewer than two columns may be used. In this example we use all eight columns. Black and white counters are placed on arbitrary squares in each column, black on one side, white on the other. (A randomizing device, such as a die, can be used for the placement.) Players sit on opposite sides and alternate moves. A move consists in advancing one of your counters any desired number of empty cells in its column. It may not leap its opposing counter, so that when two counters meet, neither may move again. The last player to move wins.

An astute reader may see at once that this game is no more than a thinly disguised nim. The "piles" are the empty cells between each pair of opposing counters. In the illustration, the piles are 5–1–4–2–0–3–6–3, which has an unsafe nim sum of

Figure 83

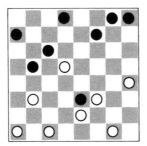

A nim game on a chessboard

4. The first player can win by moving the counter in column one, three or seven forward four spaces. If the game had begun with all the counters in each player's first row, the pattern would have been 6–6–6–6–6–6–6–6, a safe position because its nim sum is zero. The first player must lose. The second player groups the columns into four pairs, then duplicates each of his opponent's moves in the paired column, a strategy that ensures a zero nim sum after every move.

Suppose we complicate the rules by allowing either player to move backward as well as forward. Such a retreat is equivalent, of course, to *adding* counters to a nim pile. How does this affect the winning strategy?

A better-disguised game based on nim addition is a delightful pencil-and-paper game recently invented by John Horton Conway, the University of Cambridge mathematician who invented "Life," the topic of three of this book's chapters. Conway calls the new game Hackenbush, but it has also been called Graph and Chopper, Lizzie Borden's Nim and other names.

The initial pattern is a set of disconnected graphs, such as the Hackenbush Homestead as drawn by Conway [*see Figure 84*]. An "edge" is any line joining two "nodes" (spots) or one node to itself. In the latter case the edge is a "loop" (for example, each apple on the tree). Between two nodes there can be multiple edges (for example, the light bulb). Every graph stands on a base line that is not part of the graph. Nodes on the base line, which is shown as a broken line in the illustrations, are called "base nodes."

Two players alternate in removing any single edge. Gravity now enters the game because taking an edge also removes any portion of the graph that is no longer connected to the base line. For instance, removing edge *A* eliminates both the spider

Figure 84

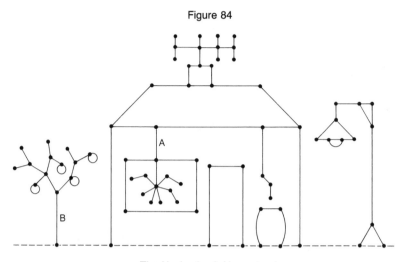

The Hackenbush Homestead

and the window since both would fall to the ground, but removing the edge joining the spider to the window removes only the spider. Taking edge B chops down the entire apple tree. If one edge of the streetlight's base is taken, the structure still stands, but taking the second edge on a later move topples the entire structure. The person who takes the picture's last edge is the winner.

As in nim, every picture is either safe (second-player win) or unsafe (first-player win), and the winner's strategy is to convert every unsafe pattern to safe. To evaluate a picture each graph must be assigned a number measuring the graph's "weight." To arrive at the assignment the first step is to collapse all the "cycles" (closed circuits of two or more edges) to loops, turning the graph into what Conway calls an apple tree, although in many cases the loops are best regarded as being flower petals. To see how it works, consider Conway's girl [see Figure 85]. She incorporates two cycles: her head and her skirt. First the two nodes of her head are brought together and then the two edges are bent into loops. Do the same with the five nodes and five edges of the skirt. The girl is now a flower girl [middle figure]. The next step is to change her to an ordinary tree by replacing each loop with a single branch [figure at right].

We now calculate this tree's weight. First, label 1 all edges with a terminal node (a node unconnected to another edge) or, to put it differently, all edges that, if removed, cause no other edges to fall off the tree. Label 2 all edges that support only

Figure 85

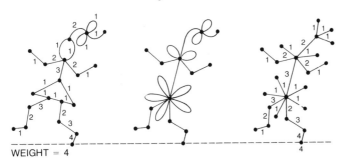

WEIGHT = 4

Girl on one foot

one edge. Each remaining edge is labeled with one more than the nim sum of all the edges it immediately supports. Consider the edge corresponding to the girl's hair between her head and her hair ribbon. It immediately supports 1–1–1. A pair of 1's cancel, giving a nim sum of 1. Add 1 to the nim sum and this edge has a weight of 2. The edge that forms the body above the skirt immediately supports edges of values 2–1–2–1–2. The nim sum is 2. Add 1 and the edge has a weight of 3.

The girl's unraised thigh supports 3–1–1–3–1–1–1, a nim sum of 1, to which 1 is added to give the thigh a value of 2. The calf below it has a value of 3, the foot a value of 4. (In each case we simply add 1 to the value of the single, immediately supported edge.) Since the foot is the only support of the entire graph, the girl has a weight of 4. All edge values are now transferred to corresponding edges on the original girl.

With practice, edge values can be computed directly on the original graph, but it requires great care. For example, the girl's five skirt edges, raised thigh, and body are all "immediately" supported by her unraised thigh. This is clear in the tree graph but is not so obvious in the original graph because many of the immediately supported edges are not close to the thigh.

If a graph has more than one base node, such as the door, barrel and lamp in the Homestead, collapse the base cycle into loops, remembering that the broken line segment between a pair of base nodes is not part of the graph. The door's transformations are shown in Figure 87b. Since the nim sum of 1–1–1 is 1, the door's weight is 1. A girl standing on both feet [*see Figure* 86] has a weight of 3. Note how the two cycles formed by her skirt and legs collapse into seven loops. A winning move, for a game played with her alone, is taking the top of her head or one of her ·hairs. This lowers the value of her

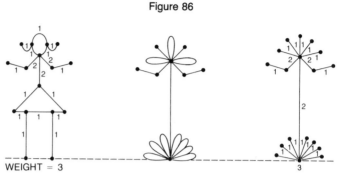

Figure 86

WEIGHT = 3

Girl on both feet

head to zero, her body to 1 and her weight to zero. In this manner a weight can be assigned to each of the five graphs that make up the Hackenbush Homestead: The apple tree, house (including window, spider, chimney, television antenna and drainpipe), door, barrel and streetlight.

If Hackenbush is played with only the girl on one foot, the game is as trivial as playing nim with only one pile. The first player can win at once by taking the supporting foot. The poor girl collapses and he acquires all her edges. In the case of a figure with more than one base node, such as the door, we must remember to take an edge so that the remaining nim sum is zero. A first player can do this only by taking the door's top edge, leaving two graphs of weight 1 each, or a combined nim sum of zero. Taking either side leaves only one graph (of weight 2), which can be taken entirely by the second player.

A picture consisting of n graphs, such as the five graphs of the Homestead, is treated exactly like five piles in nim. The nim sum of all the weights is the total Grundy number. If and only if this number is zero is the picture safe and the second player assured of winning. As in nim, the winning strategy is to play so that the nim sum of what remains is always zero.

The reader is invited to determine the weight of each graph in the Hackenbush Homestead and verify that the Homestead's nim sum is 10. Since this is not zero, the first player can win. It turns out (of course Conway designed it that way) that there is only one edge the first player can take that will guarantee a win by lowering the nim sum to zero. Which edge is it?

My account of hackenbush is only a brief introduction to this game. For a fantastic amount of additional information about the game, its deep theorems and its numerous variations, see

Conway's *On Numbers and Games,* and the two volumes of *Winning Ways* by Berlekamp, Conway, and Guy. Both works also contain an abundance of material on other nim-like games and the theory behind them in both standard and *misère* play.

ANSWERS

The first problem was to explain how the winning strategy in a chessboard version of nim is affected by allowing players to move their counters backward. The answer: It has almost no effect. If the loser retreats, the winner merely advances his opposing counter until the number of spaces separating the two men is the same as before. This preserves the status quo, leaving the basic strategy unaltered. The winner never retreats and, since the chessboard is finite, the loser's retreats must eventually cease. This variation of the game has been attributed to D. G. Northcott and is known as Northcott's nim.

How the various parts (graphs) of John Horton Conway's Hackenbush Homestead are transformed, as explained, into apple trees, then trees and labeled is shown in Figures 87, 88. The graphs have weights of 15–1–1–4–1, therefore the Homestead's nim sum is 10. The only way the first player can reduce this Grundy number to zero is by lowering the apple tree's weight to 5. "The tree trunk supports two branches of 8 and 6," Conway writes, "and these must be changed to 2 and 6, or

Figure 87 a & b

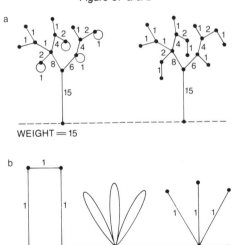

WEIGHT = 15

WEIGHT = 1

Weighing the Hackenbush apple tree, door, barrel
and streetlight

Figure 87 c & d

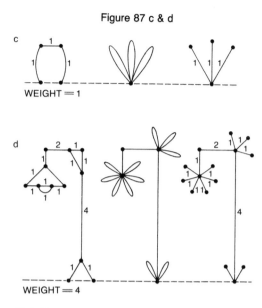

Weighing the Hackenbush apple tree, door, barrel
and streetlight

8 and 12, to have nim sum 4. Clearly we must choose the left branch. Climbing the tree, we discover that there is a unique winning move—chop the twig bearing the second apple from the left."

This chop lowers the tree's weight (the value of its trunk) to 5 [*see Figure* 89]. The graphs now have weights of 5–1–1–4–1, which have a nim sum of zero.

BIBLIOGRAPHY

"The G-Values of Various Games." Richard K. Guy and Cedric A. B. Smith. *Proceedings of the Cambridge Philosophical Society,* Vol. 52, Part 2, July 1956, pages 514–526.

"Disjunctive Games with the Last Player Losing." P. M. Grundy and C. A. B. Smith. *Proceedings of the Cambridge Philosophical Society,* Vol. 52, Part 2, July 1956, pages 527–533.

"Nim and Tac-Tix." Martin Gardner in *The Scientific American Book of Mathematical Puzzles and Diversions.* Simon and Schuster, 1959.

Nim-like Games and the Sprague-Grundy Theory. John Charles Kenyon, master of science thesis, University of Calgary, 1967.

"Compound Games with Counters." Cedric A. B. Smith. *Journal of Recreational Mathematics,* Vol. 1, April 1968, pages 67–77.

On Numbers and Games. John H. Conway. Academic Press, 1976.

Figure 88

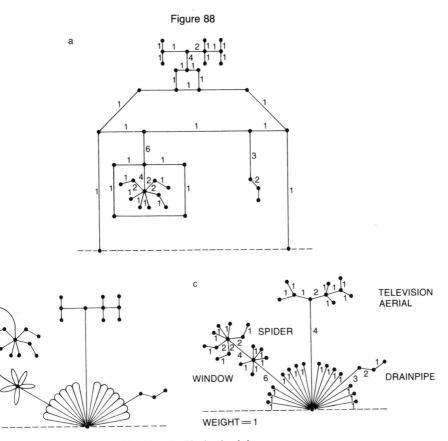

Weighing the Hackenbush house

Figure 89

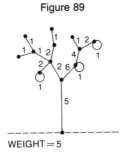

WEIGHT = 5

Apple tree after the winning chop

"Jam, Hot, and Other Games." Martin Gardner in *Mathematical Carnival*. Knopf, 1977.

Winning Ways, Vols. 1 and 2. Elwyn R. Berlekamp, John H. Conway, and Richard Guy. Academic Press, 1982.

15

GOLOMB'S GRACEFUL GRAPHS

One of the least explored areas of modern mathematics is a class of problems that combine graph theory and arithmetic. Recreational problems of this type have been discussed before in my earlier book collections; for example, in the chapter on Magic Stars and Polyhedrons in *Mathematical Carnival.* In this chapter we take up a family of numbered-graph problems that has recently been defined and developed by Solomon W. Golomb, professor of engineering and mathematics at the University of Southern California. He is the author of *Polyominoes* (Scribner's, 1965), numerous articles on recreational topics and many technical papers. What follows is extracted from his correspondence and from his paper "How to Number a Graph."

Golomb has coined the term "graceful graph" for any graph that can be "gracefully numbered." He explains this terminology with a simple example: The graceful numbering of the graph shown in Figure 90. It is called the "complete graph for four points" because every pair of its four nodes is joined by a line called an edge. The graph is topologically equivalent to the

Figure 90

A graceful graph

skeleton of a tetrahedron. It is planar because it can be drawn on the plane without intersecting edges. A graph, as we shall see, need not be planar in order to be gracefully numbered, but it must be without loops (lines joining a node to itself) or multiple edges (more than one edge connecting the same pair of nodes).

Each node is labeled with a nonnegative integer. The lowest integer (by convention) is 0, and no two integers may be alike. After the nodes are numbered every edge is labeled with the difference between the numbers of its two end nodes. Like node numbers, all edge numbers also must be distinct (no two alike). The objective is to do all these things and keep the largest node number as small as possible. Obviously it cannot be smaller than the number of edges. If the largest node number equals the number of edges, e, the edge numbers will run consecutively from 0 through e, and we shall have achieved a graceful numbering. The number e will represent three values: the total number of edges, the highest node number and the highest edge number. Any graph that can be gracefully numbered is a graceful graph. Some graceful graphs have only one basic numbering, others more than one. (Trivial variations obtained by such symmetry operations as rotations and reflections, or by replacing each node number n by $e-n$, are not considered different.) A graph that cannot be numbered gracefully is called an ungraceful graph.

As Golomb points out, every complete graph can be drawn with all its nodes on a straight line and the remaining edges can be added as curved lines [see left side of Figure 91]. Let us go further. Imagine that the straight line is the edge of a ruler with a length equal to the largest edge number of a numbered graph. The nodes of the graph are marks on the ruler at points that correspond to their numbers, each number indicating the mark's distance from the zero end of the ruler. Golomb calls such a ruler a "Euclidean model" of a numbered complete

Figure 91

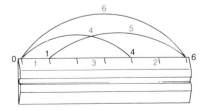 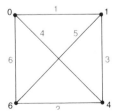

Ruler version *(left)* of a complete graph *(right)*

graph. The problem of gracefully labeling a complete graph of n nodes is equivalent to the problem of putting n marks on a ruler (always including the ruler's two ends as marks) so that every distance between a pair of marks is a distinct integer. In this example the ruler is marked at points 0, 1, 4 and 6, the node numbers of the complete graph for four points after it is gracefully numbered. Such a ruler clearly can measure lengths of one, two, three, four, five and six units. At the right of the ruler is shown another way of drawing the complete graph for four points: as a four-sided polygon with all its diagonals. (The intersection of the diagonals is not, of course, a node). Note that the distances between adjacent marks on the ruler, together with the ruler's length, correspond to the perimeter numbers of the gracefully labeled square graph.

A closely related but less restricted ruler problem was discussed in Chapter 6 of my book *The Incredible Dr. Matrix*. Dr. Matrix' rulers measure all integral distances from zero to the length of the ruler, but the numbers of its "edges" (distances between any pair of marks) are not required to be different. With the added proviso that all such distances must be different, Dr. Matrix' ruler problem becomes identical with the problem considered here: That of finding a ruler with marks that correspond to the graceful numbering of a complete graph with n nodes. Golomb proves in his paper that this can be done only if n is 1, 2, 3, or 4. Expressed differently, no complete graph for n points, when n exceeds 4, can be gracefully numbered.

If we keep the requirement that all distances between pairs of marks must be different, but we do not insist that they run consecutively from zero to the ruler's total length, we can still look for the shortest possible ruler of n marks (end points are included as marks) on which all distances between a pair of marks (which correspond to the edge numbers of the complete graph for n points) are different. In the chart of the shortest-known rulers when n is from 2 through 11 [*see Figure* 92], only the first three entries are solutions to Dr. Matrix' ruler problem. They correspond to the graceful numbering of complete graphs for two, three and four points. The other rulers do not have consecutive integral distances from zero to the ruler's length; they correspond to what Golomb calls the "best" numbering of complete graphs for more than four points. The numbers in each row give the distances between adjacent marks on rulers of two, three, four, . . . , 11 marks. The chart, which extends downward to infinity, is called the Golomb triangle.

Figure 92

NODES	EDGES	DISTANCES BETWEEN ADJACENT MARKS										LENGTH
2	1	1										1
3	3	1	2									3
4	6	1	3	2								6
5	10	1	3	5	2							11
6	15	1	3	6	5	2						17
7	21	1	3	6	8	5	2					25
8	28	1	3	6	11	8	5	2				36
9	36	1	3	12	10	8	6	5	2			47
10	45	1	3	6	12	16	11	8	5	2		64
11	55	1	8	10	5	7	21	4	2	11	3	72

GOLOMB'S TRIANGLE: Shortest Golomb rulers known in 1972.

We can put the difference between Dr. Matrix' rulers and Golomb's rulers as follows. Dr. Matrix' rulers minimize the number of marks for a ruler of length k that can measure all integral distances from 1 through k. Golomb's rulers do not necessarily include all the integral distances from 1 through k; with Golomb's rulers, for a ruler with a given number of marks, the length of the ruler is minimized and all the integral distances the ruler does measure are different. If we draw a graph corresponding to a Dr. Matrix ruler, we may find two edges with the same edge number. By omitting all edges with duplicate numbers we can get a graceful graph that Golomb calls a "graceful approximation" of a complete graph. For example, by dropping one edge (the line between points 1 and 4) from a complete graph for five points [see Figure 93] the graph can be gracefully numbered. It is equivalent to Dr. Matrix' ruler with marks at points 0, 1, 4, 7 and 9.

It is worth noting that on Golomb rulers not only are all differences between pairs of node numbers distinct, but also all

Figure 93

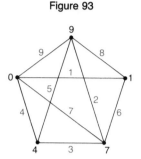

Graceful graph for a Dr. Matrix ruler

sums of pairs of node numbers, including the pairing of a node number with itself. "That this is equivalent to the differences being distinct is surprising," Golomb writes, "but fantastically simple to prove." (Proof: if $a - b = c - d$, then $a + d = b + c$, and conversely.)

With a yardstick, or 36-unit ruler as an example, here is a quick way to prove that all distances measured by a Golomb ruler are distinct. The yardstick has eight marks. The top row [*see Figure* 94], taken from Golomb's triangle, gives the distances between adjacent marks on this ruler. These seven numbers, together with the ruler's total length, correspond to the eight edge numbers on the perimeter of an eight-sided polygon when it is made into a complete graph by drawing all its diagonals and then numbered as gracefully as possible. The

Figure 94

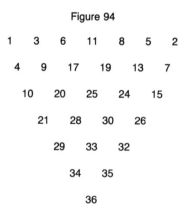

Proof for eight-mark Golomb ruler

second row of numbers is obtained by adding successive pairs of numbers in the row above it. The third row consists of adding successive triplets in the top row, the fourth row of adding successive quadruplets, and so on. The bottom number is the ruler's length. It is, of course, the sum of all the numbers in the top row. The 28 numbers of this triangle are the 28 edge numbers of the complete graph for eight points when it is given the best ungraceful numbering. If all these numbers are different, no two edge numbers of the complete graph will be alike and no two distances between pairs of marks on the corresponding ruler will be alike.

Golomb admits that for all rulers longer than six units the results were obtained (by himself and others) partly by trial and error. They have not yet been proved to be rulers of minimal length. (The ruler of length 47, for nine marks, was first found in 1965 by Matthew J. C. Hodgart of Brighton in England; the ruler of 72 lengths, for 11 marks, by Robert Reid of Miraflores in Argentina, also in 1965.) Perhaps readers can improve on these results or extend the triangle farther downward.

One of the many unusual properties for all graceful graphs discovered by Golomb is that the nodes of such graphs can always be divided into two sets—those with even numbers and those with odd—and the number of edges connecting the two sets will be $[(e+1)/2]$, where e is the total number of edges in the graph. The brackets mean that the expression is rounded down to the nearest integer. Golomb calls this a "binary labeling." For example, the even set of nodes in the graph at the left of Figure 91 are numbered 0, 4 and 6, and the odd set has only the number 1. Inspection shows that the two sets are indeed joined by $[(6+1)/2]=3$ edges.

Moreover, as Golomb proves, if all the nodes of a graph are of even order (attached to an even number of edges), the graph is graceful only if $[(e+1)/2]$ is even. When this value is odd, binary labeling is impossible and therefore the graph cannot be gracefully numbered. Of the topologically distinct graphs with five or fewer nodes, only three are ungraceful. All three have five nodes and all their nodes are of even order. The three graphs violate Golomb's condition that $[(e+1)/2]$ must be even [see Figure 95]. Note that the first two graphs are planar whereas the third, the complete graph for five points, is not. This shows that not all planar graphs, and not all nonplanar graphs, are graceful. Can a nonplanar graph be graceful? Yes, as the graceful labeling of the Thomsen graph shows [see Figure 96]. The Thomsen graph is sometimes called the utilities graph because it diagrams the well-known (and unsolv-

Figure 95

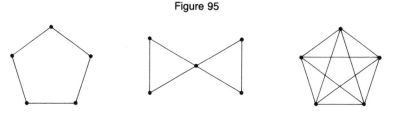

The only ungraceful graphs with fewer than six nodes

Figure 96

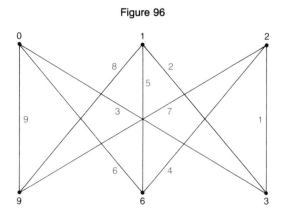

A graceful numbering of the Thomsen graph

able) puzzle in which three houses are each to be connected to three utilities without any crossing of edges. The Thomsen graph is one of an infinite family of graphs, known as "complete bipartite graphs," in which every node in a set of a nodes is joined to every node in a set of b nodes, but nodes within each set are not connected. Golomb has established that all complete bipartite graphs are graceful.

Skeletons of polyhedrons can be represented as planar graphs known as Schlegel diagrams. Of the five Platonic solids only the dodecahedron and icosahedron have not been shown to be graceful. We have seen how to gracefully number the tetrahedron. Can the reader gracefully number the Schlegel diagrams of the cube and octahedron [*see Figure* 97] before Golomb's labelings are given in the Answer Section? Can he do the same for the diagram of the skeleton of the Great Pyramid of Egypt? Can he discover graceful numberings for the dodecahedron or the icosahedron?

Three other graceful graphs by Golomb have six, seven and

Figure 97

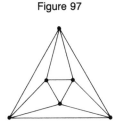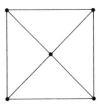

Three graceful Schlegel graphs: cube *(left)*,
octahedron *(center)* and Great Pyramid *(right)*

Figure 98

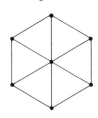

Three graceful graphs by Golomb with six, seven,
and ten nodes

10 nodes [*see Figure* 98]. Can the reader number these also before the solutions are given?

In addition to complete bipartite graphs there are other infinite families of graceful graphs. One found by Golomb is
shown in Figure 99. The question arises: As the number of
nodes approaches infinity, does the fraction of graceful graphs
among all graphs of n nodes approach a limit? If so, what is
the limit? For several years no fractional value from 0 through

Figure 99

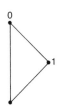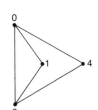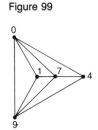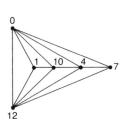

An infinite family of graceful graphs

1 was excluded, but recently Paul Erdös has been able to show
that the limit is 0. His proof, not yet published, is difficult.
Gary Bloom and Herbert Taylor found a fairly easy way to
show that the number of graceful graphs with e edges is
equal to or less then e, from which it follows at once that the
limit is 0.

Although many unsolved problems about graceful graphs,
some very technical, have now been cleared up by Golomb,
Erdös, and others, there are still several major questions that
remain unanswered:

(1) What are the necessary and sufficient conditions for a
graph to be graceful? It is not even known if all tree graphs
are graceful. (Tree graphs are discussed in Chapter 17 of my
Mathematical Magic Show.) Gerhard Ringel in 1963 apparently
was the first to conjecture, in a different terminology and independently of Golomb's work, that all tree graphs can be
gracefully numbered. This has been the subject of several papers by Alexander Rosa and other Czechoslovakian mathematicians. The conjecture has been established only for special
kinds of trees such as "caterpillars"; trees with every node on
a central stalk or only one edge from the stalk [*see Figure* 100].
In a typical gracefully numbered caterpillar the edge numbers
run consecutively from one end of the tree to the other.

Figure 100

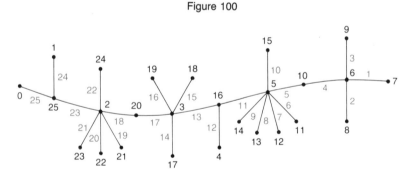

A graceful caterpillar

Golomb has discovered a similar algorithm for gracefully
numbering an infinite class of polyomino graphs such as the
pentomino and the heptomino [*see Figure* 101]. Note how the
consecutive numbers run diagonally upward, from left to right.
Unfortunately there is an infinite class of polyominoes with a
greater degree of concavity (the degree is not easy to define)

Figure 101

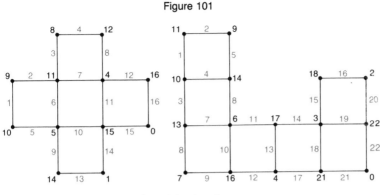

Graceful polyominoes

for which this procedure fails even when they can be gracefully numbered.

A simple graph found by Golomb [*see Figure* 102] is particularly ungraceful because it is not ruled out by any known general theorem.

Figure 102

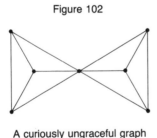

A curiously ungraceful graph

(2) What are the rules for forming Golomb's triangle? Put another way, is there a general algorithm for finding the shortest rulers that correspond to the best ungraceful numbering of a complete graph for more than four points?

(3) Is there a graph that, when numbered as gracefully as possible, violates the conjecture that on all such graphs the highest node number and the highest edge number are equal? Golomb is now searching for a counterexample; a graph with the best numbering but with a highest node number that exceeds the highest edge number. (It cannot be the other way around.) "If I find one," Golomb writes in a letter, "the graph will not only be ungraceful but downright *disgraceful*."

Figure 103

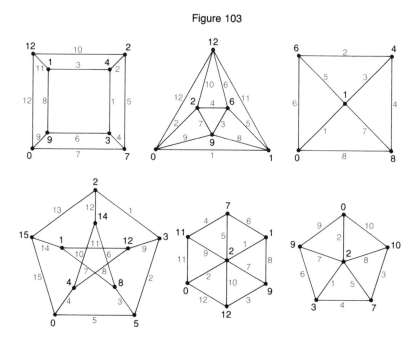

Solutions to the graceful-graph problems

ANSWERS

Solutions to the six graphs that readers were asked to number "gracefully" are shown in Figure 103. None of these numberings is unique.

Readers were also asked to improve on the rows of "Golomb's triangle," each row giving the shortest-known rulers of n marks (including end points) such that every distance between a pair of marks is a distinct integer. Walter Penney of Greenbelt, Md., was the first to lower the eight-mark ruler to length 34. The same ruler was also found by hand by Daniel A. Lynch of Wildwood, N.J.

William Mixon of the University of Chicago was the first to make an exhaustive computer search for all minimum-length rulers through 11 marks. His results show that rulers of eight, nine and 10 marks are unique, except of course for reversals [*see Figure* 104]. These results were completely confirmed by Ashok Kumar Chandra's computer program at Stanford University and partly confirmed by the programs of Paul Steier, James R. Van Zandt, Edward Schonberg and others. Working by hand, Sheldon B. Akers found the nine-mark ruler, and

Figure 104

NODES	LENGTH	DIVISIONS
3	3	1, 2
4	6	1, 3, 2
5	11	1, 3, 5, 2
		2, 5, 1, 3
6	17	1, 3, 6, 2, 5
		1, 3, 6, 5, 2
		1, 7, 3, 2, 4
		1, 7, 4, 2, 3
7	25	1, 3, 6, 8, 5, 2
		1, 6, 4, 9, 3, 2
		1, 10, 5, 3, 4, 2
		2, 1, 7, 6, 5, 4
		2, 5, 6, 8, 1, 3
8	34	1, 3, 5, 6, 7, 10, 2
9	44	1, 4, 7, 13, 2, 8, 6, 3
10	55	1, 5, 4, 13, 3, 8, 7, 12, 2
11	72	1, 3, 9, 15, 5, 14, 7, 10, 6, 2
		1, 8, 10, 5, 7, 21, 4, 2, 11, 3

Minimum-length Golomb rulers

Wolfgang Harries, also working by hand, found all but one of the rulers with six and seven marks.

The 10-mark ruler and one 11-mark ruler had been found earlier by John P. Robinson of the University of Iowa with a nonexhaustive computer search made in connection with work on his 1966 doctorate on error-correcting codes. His best results, for rulers through 24 marks, are given in "A Class of Binary Recurrent Codes with Limited Error Propagation," by Robinson and Arthur J. Bernstein, in *IEEE Transactions on Information Theory* (Volume IT-13, Number 1, January, 1967, pages 106–113).

R. C. Ashenfelter of the Bell Telephone Laboratories was the first to gracefully label the dodecahedron. Chandra devised a computer program that made an exhaustive search for the icosahedron and produced five fundamentally different labelings. A partial search for the dodecahedron yielded a large number of graceful labelings. This settles affirmatively Golomb's conjecture that the skeletons of all five Platonic solids are graceful graphs.

ADDENDUM

Many early references to Golomb rulers, in other terminologies, have come to light. The earliest known to me is a problem by Henry D. Friedman that appeared in the *SIAM Review*, Vol. 5, July 1963, page 275.

Golomb rulers have practical applications to pulsed radar and sonar codes (see "Synch-Sets: A Variant of Difference Sets," by G. J. Simmons, *Proceedings of the Fifth Southeastern Conference on Combinatorics, Graph Theory and Computing*, Boca Raton, 1974, pages 625–645) and to X-ray diffraction crystallography. Two Golomb rulers of length 17 provide counterexamples to a "theorem" published by S. Picard in 1939 and used in crystallography for many years.

Richard Guy reported (*The American Mathematical Monthly*, Vol. 88, December 1981, page 756) that since Golomb revived Ringel's conjecture that all tree graphs are graceful, some 100 papers have dealt with partial results on this notorious and still unanswered question.

Ronald L. Graham and Neil Sloane, both of Bell Laboratories, have defined a "harmonious graph" as follows: A connected graph, with n edges, is harmonious if its points can be labeled with distinct integers (modulo n) so that the sums of the pairs of numbers at the ends of each edge are also distinct (modulo n). Harmonious graphs have much in common with graceful graphs, and are related to error-correcting codes and to a famous combinatorial problem known as the postage stamp problem. See "On Additive Bases and Harmonious Graphs," by Graham and Sloane, *SIAM Journal on Algebraic and Discrete Methods*, Vol. 1, December 1980, pages 382–404. The authors show (among many other things) that graphs known as ladders, fans, and wheels are harmonious. Trees (with zero repeated once) may be harmonious. The Petersen graph and skeletons of the tetrahedron, dodecahedron, and icosahedron are harmonious. Skeletons of the cube and octahedron are not. Almost all graphs, the authors conclude, are neither harmonious nor graceful.

In recent years Golomb and Herbert Taylor have been exploring a two-dimensional analog of ruler problems which have many practical applications. See their paper on "Two-dimensional Synchronization Patterns for Minimum Ambiguity," in *IEEE Transactions on Information Theory*, Vol. IT–28, July 1982, pages 600–604, and Golomb's "Algebraic Constructions for Costas Arrays," to appear in *Journal of Combinatorial Theory*, Series A, sometime in 1983.

BIBLIOGRAPHY

"How to Number a Graph." S. W. Golomb in *Graph Theory and Computing*. Ronald C. Read (ed.) Academic Press, 1972.

"The Largest Graceful Subgraph of the Complete Graph." S. W. Golomb. *The American Mathematical Monthly*, Vol. 81, May 1974, pages 499–501. See also comments by Richard Guy in the same journal, Vol. 82, December 1975, page 1,000.

"Miami Beach." Martin Gardner in *The Incredible Dr. Matrix*. Scribner's, 1976.

Applications of Numbered Undirected Graphs. Gary S. Bloom and Solomon W. Golomb. *Proceedings of the Institute of Electrical and Electronics Engineers*, Vol. 65, April 1977, pages 562–570. (G. S. Bloom and S. Golomb are anagrams, but they are distinct mathematicians.)

"Graceful Graphs, Radio Antennae, and French Windmills." J. C. Bermond in *Graph Theory and Combinatorics*. R. J. Wilson (ed.) London: Pitman, 1979.

"The Sparse Ruler." Problem 1076 in *Journal of Recreational Mathematics*, Vol. 15, No. 2, 1982–83, pages 152–155.

16

CHARLES ADDAMS' SKIER

AND OTHER PROBLEMS

1. THE FLEXIBLE BAND

Gustavus J. Simmons, in charge of research and development at Rolamite Inc., Albuquerque, N.M., sent this curious topological problem. Work at Rolamite involves complex banded rolling systems. One of the Rolamite engineers, Virgil Erbert, was confronted in the course of his work with the problem shown in Figure 105. End *A* of a flexible band was fastened to an object that was too large to pass through the slot at end *B*. It was essential that the band be formed into the looped·configuration shown in the illustration without detaching end *A* from the object to which it was fastened. Can it be done?

It looks impossible, but the answer is yes. The reader is invited to draw a rough facsimile of the band on a sheet of pa-

Figure 105

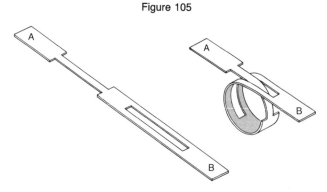

A looped-band topological puzzle

per, cut it out and tape end *A* to a tabletop. The puzzle, which is not difficult, is to manipulate the strip into the looped configuration.

2. THE ROTATING DISK

Six players—call them *A*, *B*, *C*, *D*, *E* and *F*—sit around a circular table divided into six equal parts. At the center of the table is a disk mounted on a central pin around which it can rotate [*see Figure* 106]. The disk is marked with arrows and digits.

Figure 106

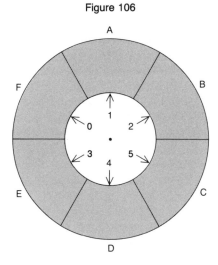

D. St. P. Barnard's game problem

The wheel is spun five times. After each spin each player scores the number of points within his segment of the table. (If the wheel stops with its arrows exactly between adjacent players, the spin is not counted.) The players keep a running total of points, and the one with the largest total after the fifth spin is the winner. If there are ties for the highest score, no one wins and the game is played again.

The outcome of the first spin is shown in the illustration. *C* is ahead with five points. After the second spin *D* is ahead. After the fifth spin *A* is the winner. What was each player's final score? The information seems to be insufficient, yet the question can be answered accurately by deductive reasoning. This unusual logic problem is adapted from a puzzle in one of D. St. P. Barnard's popular "Brain-twister" columns in the British *Observer*.

3. FRIEZE PATTERNS

A frieze is a pattern that endlessly repeats itself along an infinite strip. Such patterns can exhibit different kinds of basic symmetry, but here we shall be concerned only with what is called "glide symmetry." A glide consists of a slide (more technically a "translation") combined with mirror reflection and a half-turn. For example, repeatedly gliding the letter R to the right along a strip generates the following frieze:

RᴚRᴚRᴚRᴚRᴚRᴚRᴚR...

H. S. M. Coxeter, a geometer at the University of Toronto, recently investigated in depth a remarkable class of frieze patterns that can be constructed very simply by using nonnegative integers, if the lack of symmetry in the shapes of the numerals is ignored [*see Figure* 107]. Think of the numerals as representing spots of colors, all 1's the same color, all 2's another color and so on. In this instance any rectangular portion of the frieze that is nine columns wide, such as the shaded one shown here, can be regarded as the unit pattern. By gliding it left or right—that is, sliding and simultaneously reflecting and inverting—the infinite frieze pattern is generated.

Figure 107

• • •	0		0	0	0	0	0	0	0	0		0	• • •
• • •		1	1	1	1	1	1	1	1	1		• • •
• • •			1	2	5	2	1	3	2	2		• • •
• • •			1	9	9	1	2	5	3			• • •
• • •			1	4	16	4	1	3	7	7		• • •
• • •		1	3	7	7	3	1	4	16			• • •
• • •	1	2	5	3	5	2	1	9	9			• • •
• • •		1	3	2	2	3	1	2	5			• • •
• • •			1	1	1	1	1	1	1	1		• • •
• • •	0		0	0	0	0	0	0	0	0		• • •

A frieze pattern with glide symmetry

To produce this type of frieze pattern, begin with infinite borders of 0's and 1's at top and bottom, and a path of numbers from top to bottom such as the zig-zag path of eight 1's shown on the left between the borders of 0's. The numbers in such a path (which may be straight, or crooked as it is here), as well as the length of the path, can be varied to produce different patterns. A simple formation rule, common to all such patterns, is now applied to obtain all the other integers. The surprising glide symmetry that results is a nontrivial consequence of this rule.

Our puzzle, suggested by Coxeter, is to guess the simple rule. Hint: It can be written as an equation with three terms involving nothing more than multiplication and addition, and no exponents. When Coxeter first showed the pattern given here to the mathematician Paul Erdös, Erdös guessed the rule in 20 seconds.

A discussion of the properties of such friezes, their fascinating historical background and their applications to determinants, continued fractions and geometry can be found in Coxeter's "Frieze Patterns" in *Acta Arithmetica*, Volume 18 (1971), pages 297–310. On friezes in general and their seven basic kinds of symmetry see Coxeter's modern classic, *Introduction to Geometry* (Wiley, 1961), pages 47–49.

4. THE CAN OF BEER

On a picnic not long ago Walter van B. Roberts of Princeton, N.J., was handed a freshly opened can of beer. "I started to put it down," he writes, "but the ground was not level and I thought it would be well to drink some of the beer first in order to lower the center of gravity. Since the can is cylindrical, obviously the center of gravity is at the center of a full can and will go down as the beer level is decreased. When the can is empty, however, the center of gravity is back at the center. There must therefore be a point at which the center of gravity is lowest."

Knowing the weight of an empty can and its weight when filled, how can one determine what level of beer in an upright can will move the center of gravity to its lowest possible point? When Roberts and his friends worked on this problem, they found themselves involved with calculus: Expressing the height of the center of gravity as a function of the height of the beer, differentiating, equating to zero and solving for the minimum value of the height of the center of gravity. Later Roberts thought of an easy way to solve the problem without calculus. Indeed, the solution is simple enough to get in one's head.

To devise a precise problem assume that the empty can weighs 1½ ounces. It is a perfect cylinder and any asymmetry introduced by punching holes in the top is disregarded. The can holds 12 ounces of beer, therefore its total weight, when filled, is 13½ ounces. The can is eight inches high. Without using calculus determine the level of the beer at which the center of gravity is at its lowest point.

5. THE THREE COINS

Three coins are on the table; a quarter, a half-dollar and a silver dollar. Smith owns one coin and Jones owns the other two. All three coins are tossed simultaneously.

It is agreed that any coin falling tails counts zero for its owner. Any coin falling heads counts its value in cents. The tosser who gets the larger score wins all three coins. If all three come up tails, no one wins and the toss is repeated.

What coin should Smith own so that the game is fair, that is, so that the expected monetary win for each player is zero?

David L. Silverman, author of the excellent book of game puzzles called *Your Move* (McGraw-Hill, 1971), is responsible for this new and unpublished problem. It has an amazing answer. Even more astonishing is a generalization, formally proved by Benjamin L. Schwartz, of which this problem is a special case.

6. KOBON TRIANGLES

Kobon Fujimura, a Japanese puzzle expert, recently invented a problem in combinatorial geometry. It is simple to state, but no general solution has yet been found. What is the largest number of nonoverlapping triangles that can be produced by n straight line segments?

It is not hard to discover by trial and error that for $n = 3$, 4, 5 and 6 the maximum number of triangles is respectively one, two, five and seven [*see Figure* 108]. For seven lines the problem is no longer easy. The reader is asked to search for the maximum number of nonoverlapping triangles that can be produced by seven, eight and nine lines.

The problem of finding a formula for the maximum number of triangles as a function of the number of lines appears to be extremely difficult.

Figure 108

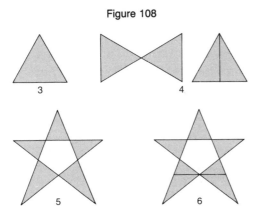

Maximum number of nonoverlapping triangles for
three, four, five, and six lines

7. A NINE-DIGIT PROBLEM

One of the satisfactions of recreational mathematics comes
from finding better solutions for problems thought to have
been already solved in the best possible way. Consider the following digital problem that appears as Number 81 in Henry
Ernest Dudeney's *Amusements in Mathematics*. (There is a Dover
reprint of this 1917 book.) Nine digits (0 is excluded) are arranged in two groups. On the left a three-digit number is to be
multiplied by a two-digit number. On the right both numbers
have two digits each:

$$\begin{array}{cc} 158 & 79 \\ \underline{23} & \underline{46} \end{array}$$

In each case the product is the same: 3,634. How, Dudeney
asked, can the same nine digits be arranged in the same pattern to produce as large a product as possible, and a product
that is identical in both cases? Dudeney's answer, which he said
"is not to be found without the exercise of some judgment and
patience," was

$$\begin{array}{cc} 174 & 96 \\ \underline{32} & \underline{58} \\ 5{,}568 & 5{,}568 \end{array}$$

Victor Meally of Dublin County in Ireland later greatly improved on Dudeney's answer with

$$\begin{array}{cc} 584 & 96 \\ \underline{12} & \underline{73} \\ 7,008 & 7,008 \end{array}$$

This remained the record until last year, when a Japanese friend of Fujimura's found an even better solution. It is believed, although it has not yet been proved, to give the highest possible product. Can the reader find it without the aid of a computer?

8. CROWNING THE CHECKERS

A well-known problem with checkers is begun by placing eight checkers in a row. A move consists in picking up a checker, carrying it right or left over exactly two checkers, then placing it on a checker to make a king. (Carrying a checker over a king counts as moving it over two checkers.) In four moves form four kings. The problem is not difficult, and it is easy to show that for any even number of checkers, n, when n is at least 8, a row of $n/2$ kings can always be produced in $n/2$ moves.

Numerous variants on this old problem have been proposed by Dudeney and other puzzle inventors. The following variation on the theme, which I believe is new, was suggested and solved by W. Lloyd Milligan of Columbia, S.C.

An even number of checkers, n, are placed in a row. First move a checker over one checker to make a king, then move a checker over two checkers, then a checker over three checkers, and so on, each time increasing by one the number of checkers to be passed over. The objective is to form $n/2$ kings in $n/2$ moves.

Can the reader prove that the problem cannot be solved unless n is a multiple of 4, and give a simple algorithm (procedure) for obtaining a solution in all cases where n is a multiple of 4? A solution is easily found by trial and error when n is 4 or 8, but for $n = 16$ or higher it is not so easy without a systematic method.

9. CHARLES ADDAMS' SKIER

Single-panel gag cartoons, like Irish bulls, are often based on outrageous logical or physical impossibilities. Lewis Carroll liked to tell about a man who had such big feet that he had to put his pants on over his head. Almost the same kind of impossibility is the basis of a famous *New Yorker* cartoon by Charles Addams of a woman skier going down a slope. Behind her you see her parallel ski tracks approaching a tree, going around the tree with a track on each side and then becoming parallel again.

Suppose you came on a pair of such ski tracks on a snowy slope, going around a tree exactly as in Addams' cartoon. Assume that they are, in truth, tracks made by skis. Can you think of at least six explanations that are physically possible?

ANSWERS

1. How to form a loop with the Rolamite band while end *A* is taped to a tabletop is shown in Figure 109.

Figure 109

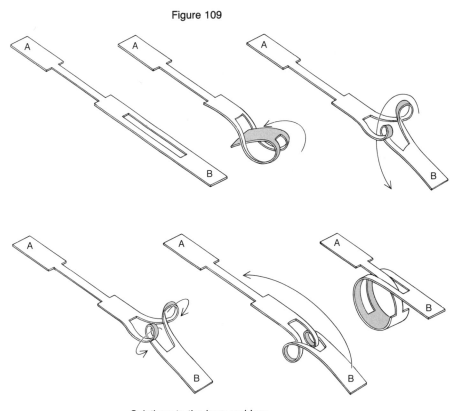

Solutions to the loop problem

Robert Neale, whom we encountered in the chapter on paper folding, suggested applying this to a playing card, say the joker. Use a razor blade to cut along the lines shown on the card at the left of Figure 110. Discard the shaded cut-out re-

Figure 110

The curious joker

gion. By carefully executing the trick bend with the little square loop, taking care not to crease or tear its sides, you can produce the structure shown on the right. It is an amusing curiosity to carry in a wallet and show to friends. How the devil was it made? It looks, of course, as if the entire card had to be somehow pushed through the tiny window!

2. The first two rows of the chart [*see Figure* 111] show the results of the first two spins. The first spin was given, and we were told that *D* had the highest total after the second spin. This could happen only if the wheel distributed the points as shown in the second row. Now comes the tricky part. Every opposite pair of digits on the disk used in the game add to 5. This

Figure 111

SPINS	A	B	C	D	E	F
1	1	2	5	4	3	0
2	0	1	2	5	4	3
3	5	4	3	0	1	2
4	5	4	3	0	1	2
5	4	3	0	1	2	5
FINAL SCORES	15	14	13	10	11	12

Solution to D. St. P. Barnard's problem

means that every spin will give a combined sum of five points to each pair of players seated opposite each other—namely *AD*, *BE* and *CF.* At the end of the game, which has five spins, each of these pairs of players will have a combined sum of 25 points.

We know that *A* won the game. Since his was the highest score, *D* (who sits opposite) must have ended with the lowest score. *D*'s final score must be less than 13, otherwise *A*'s final score would be smaller. *D*'s final score cannot be 12. True, *A* would score 13, but then a player of the pair *BE*, as well as a player of the pair *CF*, would necessarily score 13 or better, preventing *A* from being the highest scorer.

As we have seen, *D* cannot score more than 11. He already has nine points at the end of the second spin, threfore, at least one of the three remaining spins must give him zero. Since the order of the results of each spin cannot affect the final scores, we can assume that *D* scored zero after the third spin. This determines the points for the other players as indicated in the third row of the chart.

On the next two spins, *D*'s points can only be 0–0, 0–1, 1–1 or 0–2. We test each in turn. If 0–0 or 0–2, *A* will tie with someone on his final score. If 1–1, *F* gets 5–5 and wins with a score of 15. Only 0–1 remains for *D*. This makes *A* the winner, with 15 points, and enables us to complete the chart as indicated. We do not know the order of the last three spins, but the final scores are accurate. The problem is No. 2 in D. St. P. Barnard's first puzzle book, *Fifty Observer Brain-Twisters* (Faber and Faber, Ltd., 1962).

3. The formation rule for H. S. M. Coxeter's frieze patterns is that every four adjacent numbers

$$b$$
$$a \quad d$$
$$c$$

satisfy the equation $ad = bc + 1$.

4. Walter van B. Roberts answered his beer-can problem this way: "Imagine that the beer is frozen so that the can of beer can be placed horizontally on a knife-edge pivot and balanced with the can's top to the left. If it balances with the pivot under the beer-filled part, adding more beer would make the can tip to the left, whereas removing beer would make it tip to the right. If it balances with the pivot under the empty part, the reverse would be true. But if it balances with the pivot exactly under the beer's surface, any change in the amount of beer will make the can tip to the left [*see Figure* 112]. Since in this case the center of gravity moves toward the can's top when any

Figure 112

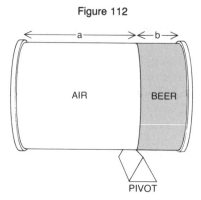

AIR   BEER

PIVOT

The balanced beer can

change is made in the amount of beer, the center of gravity must be at its lowest point when it coincides with the beer's surface.

"With the can balanced in this condition, imagine that the ends are removed and their mass distributed over the side of the can. This cannot upset the balance because it does not shift the center of gravity of the system, but it allows us to consider the can as an open-ended pipe whose mass per unit length on the empty (left) side is proportional to the weight of an empty can, whereas the mass on the beer-filled right side is proportional to the weight of a full can. The moment of force on the left is therefore proportional to the weight of an empty can multiplied by the square of the length of the empty left side, and the moment on the right side is similarly proportional to the weight of a full can multiplied by the square of the length of the beer-filled right side. Since the can is balanced, these moments must be equal.

"Pencil and paper are now hardly required to deduce that the square of the length of the empty part divided by the square of the length of the full part equals the weight of a full can divided by the weight of an empty can, or, finally, that the ratio of the length of the empty part to the full part is the square root of the ratio of the weight of a full can to an empty one."

Expressed algebraically, let a and b stand for the lengths of the empty and filled parts of the can when the center of gravity is at its lowest point and E and F for the can's weight when empty and full. Then $a^2E = b^2F$, or $a/b = \sqrt{F/E}$.

In the example given, the can weighs nine times as much when it is full as it does when it is empty. Therefore the center of gravity reaches its lowest point when the empty part is three times the length of the full part, in other words, when the beer fills the can's lower fourth. Since the can is eight inches high, the level of the beer is $8/4 = 2$ inches.

After the above solution appeared in *Scientific American,* Mark H. Johnson wrote to say that the answer is not strictly accurate. Because the tops and bottoms of the frozen can are at unequal distances from the pivot, they exert unequal moments of force. Distributing their masses over the can's side, to make a uniform and open pipe, would tilt the can slightly to the air side. To solve the problem precisely one needs more data about the can's dimensions and the masses of its top, bottom and side. Other readers reported that the solution also neglects what naval architects and engineers call the "free surface effect." When liquids are free to move inside containers, a slight raising of the vessel's center of gravity results.

5. Regardless of which coin Smith chooses, the game is fair. The payoff matrices show [*see Figure* 113] that in every case the person least likely to win (because he has only one coin) wins just enough when he does win to make both his expectation and that of his opponent zero.

As David Silverman suspected when he found this solution, the problem is a special case of the following generalization. If

Figure 113

	75	50	25	0	
WIN	75	75	75	75	SUM = 300
LOSS	− 100	− 100	− 100	0	SUM = 300

	125	100	25	0	
WIN	− 50	− 50	125	125	SUM = 150
LOSS	− 50	− 50	− 50	0	SUM = 150

	150	100	50	0	
WIN	− 25	− 25	− 25	150	SUM = 75
LOSS	− 25	− 25	− 25	0	SUM = 75

Payoffs for player with silver dollar *(top,)* half-dollar *(middle)*, and quarter *(bottom)*

a set of coins have values that are adjacent in the doubling series 1—2—4—8—16 . . . and the game is played as described, it is a fair game regardless of how the coins are divided. We assume, of course, that each player has at least one coin and that each value is represented by only one coin.

Daniel S. Fisher, a high school student in Ithaca, N.Y., generalized Silverman's generalization. He showed that Silverman's game is fair for any division of ownership of the coins when values of the coins are 1, n, n^2, . . . , n^k and the coins are weighted to fall tails with probability $1/n$ (n equal to or greater than 2).

6. The maximum number of nonoverlapping triangles that can be produced by seven, eight and nine lines are 11, 15 and 21 respectively [*see Figure* 114]. These are thought, although not yet proved, to be maximal solutions.

Figure 114

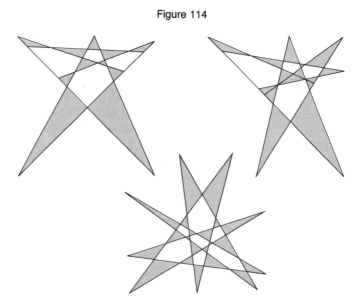

Solution to triangle problem

7. The solution with the largest product is:

$$
\begin{array}{r}
532 \\
\underline{14} \\
\hline 7{,}448
\end{array}
\qquad
\begin{array}{r}
98 \\
\underline{76} \\
\hline 7{,}448
\end{array}
$$

The problem has 11 basic solutions:

$$532 \times 14 = 98 \times 76 = 7{,}448$$
$$584 \times 12 = 96 \times 73 = 7{,}008$$
$$174 \times 32 = 96 \times 58 = 5{,}568$$
$$158 \times 32 = 79 \times 64 = 5{,}056$$
$$186 \times 27 = 93 \times 54 = 5{,}022$$
$$259 \times 18 = 74 \times 63 = 4{,}662$$
$$146 \times 29 = 73 \times 58 = 4{,}234$$
$$174 \times 23 = 69 \times 58 = 4{,}002$$
$$134 \times 29 = 67 \times 58 = 3{,}886$$
$$138 \times 27 = 69 \times 54 = 3{,}726$$
$$158 \times 23 = 79 \times 46 = 3{,}634$$

Many readers found all eleven by hand, others found them with computer programs. Allan L. Sluizer pointed out that the maximum answer has digits 1 through 5 in one of the multiplications, and digits 6 through 9 in the other.

If 0 is included among the digits (though not as an initial digit of a number), we may ask for solutions of the expression $abc \times de = fgh \times ij$. There are 64 solutions, all independently found by Richard Hendrickson, R. F. Forker, and Sluizer. The one with the smallest product is $306 \times 27 = 459 \times 18 = 8{,}262$. The one with the largest product is $915 \times 64 = 732 \times 80 = 58{,}560$. The maximum solution is given by Dudeney, in his answer to problem 82 of *Amusements in Mathematics*, as the maximum product obtainable if the ten digits are divided in any manner whatever to form a pair of multiplications, each of which gives the product. As in all such problems, 0 may not be an initial digit. The lowest product is given by $3{,}485 \times 2 = 6{,}970 \times 1 = 6{,}970$.

"It is extraordinary," Dudeney once declared, "what a large number of good puzzles can be made out of the ten digits." Here are some examples similar to our original problem. How many solutions are there to $ab \times cde = fghi$, using the nine positive digits? And how many to $a \times bcde = fghi$? The seven solutions to the first problem, and the two to the second, are given by Dudeney in his answer to problem 80, *Amusements in Mathematics*.

Using all ten digits, how many solutions are there for $ab \times cde = fghij$? I have not seen this answered in print, but Y. K. Bhat, a correspondent in New Delhi, found nine:

$$39 \times 402 = 15{,}678$$
$$27 \times 594 = 16{,}038$$
$$54 \times 297 = 16{,}038$$
$$36 \times 495 = 17{,}820$$
$$45 \times 396 = 17{,}820$$
$$52 \times 367 = 19{,}084$$
$$78 \times 345 = 26{,}910$$
$$46 \times 715 = 32{,}890$$
$$63 \times 927 = 58{,}401$$

How about $ab \times c = de + fg = hi$, excluding 0? In *Modern Puzzles*, problem 73, Dudeney gives the only answer: $17 \times 4 = 93 - 25 = 68$.

Clement Wood, in his rare *Book of Mathematical Oddities* (Little Blue Book No. 1210), asserts that $ab \times c = de \times f = ghi$ (0 excluded) has only two solutions: $38 \times 4 = 78 \times 2 = 156$, and $58 \times 3 = 29 \times 6 = 174$.

One final problem that I leave unanswered. Find the only solution (excluding 0) to $a \times bc = d \times ef = g \times hi$. This was sent to me in 1972 by Guy J. Crocker, who discovered it. I cannot recall having seen it before.

8. If n is odd, it is obvious there is no solution. If n is even but not a multiple of 4, an odd number of checkers must be jumped on the final move. This would necessarily leave a single checker in the row, therefore the assumption that there is a solution when n is not a multiple of 4 must be false.

If there are $4n$ checkers, the problem can be solved by working it backward according to the following algorithm. Start with $n/2$ kings in a row. Take the top checker from either of the two middle kings, jump over the largest group of kings and put down the checker as a single man. On the next backward move take the top checker from the other middle king and jump in the same direction as before, jumping one fewer checker. Follow this procedure until all kings in the direction of the first jump are eliminated. Take the top checker from the inside king and jump in the same direction as the previous jumps, moving it over the proper number of checkers. Continue this procedure, always in the same direction, until all the kings are reduced to single men. When these moves are taken in reverse order, they provide one solution (there are many others) to the original problem.

9. Here are six possible explanations of the ski tracks:

(1) The skier bumped into the tree but protected himself with his hands. Keeping one ski in place, he carefully lifted his other foot and moved to the lower side of the tree. With his back against the tree, he replaced his raised foot and ski on the other side, then continued down the slope.

(2) The skier slammed into the tree with such force that his skis came off and continued down the slope without him.

(3) Two skiers went down the hill, each wearing only one ski.

(4) One skier went down the hill twice, each time with one ski on one foot.

(5) A skier went down a treeless slope, moving his legs apart at one spot. Shortly thereafter a tree with a sharpened trunk base was plunged into the snow at that spot.

(6) The skier wore stilts that were high enough and sufficiently bowed to allow him to pass completely over the tree.

So many readers sent other preposterous explanations that I can give only a sampling:

A small, supple tree that bent as the skier went over it was proposed by John Ferguson, John Ritter, Brad Schaefer, Oliver G. Selfridge and James Weaver. Ferguson also suggested (among his 23 possibilities) a pair of skis pulled uphill by long ropes and two toboggan teams of *very* small midgets, four on each ski. Selfridge included this one: The skier, aware of his ineptness, wore a protective lead suit. His impact on the tree sheared out a cylindrical section. The dazed skier passed between top and bottom parts of the tree before the top fell down and balanced perfectly on the base.

Manfred R. Schroeder, director of the Drittes Physikalisches Institut at the University of Göttingen, reported an actual experience he had in 1955 while skiing down a mountain in New Hampshire. "I hit a small but sturdy tree with my right shinbone. The binding came loose and the ski and leg went around different sides of the tree. Below the tree, leg and ski came together again. However, the binding did not engage (no automatic step-in bindings then!) and the tracks ended in a spill about ten yards farther down the slope. Even then, in spite of considerable pain in the leg, I thought it was a worthwhile experience."

Johnny Hart, in his *B.C.* comic strip, has played with the theme. Thor, speeding toward a tree on his stone wheel unicycle, once went around the tree leaving two tracks. What happened on a later occasion is reproduced in Figure 115.

Figure 115

By permission of Johnny Hart and Field Enterprises, Inc.

BIBLIOGRAPHY

For more on the beautiful properties of frieze patterns see:

"Triangulated Polygons and Frieze Patterns." J. H. Conway and H. S. M. Coxeter. *Mathematical Gazette,* Vol. 57, October 1973, pages 87–94.

"Additive Frieze Patterns and Multiplication Tables." G. C. Shephard. *Mathematical Gazette,* Vol. 60, October 1976, pages 178–184.

17

CHESS TASKS

Everyone who calls a [chess] problem
"beautiful" is applauding mathematical beauty,
even if it is beauty of a comparatively lowly
kind. Chess problems are the hymn-tunes of
mathematics.

—G. H. HARDY, *A Mathematician's Apology*

It has been my policy to avoid chess problems of the type
"Mate in *n* moves" on the assumption (perhaps a mistaken one)
that too few readers play chess and that, even among those
who do, too few like chess problems. In this chapter, however,
I shall consider a variety of what are called chess "task" problems. They have so little in common with actual play that they
are of more interest to puzzle buffs than to serious chess players. True, a knowledge of chess rules is essential. But apart
from that, even a tyro is as likely as a grandmaster to be able
to solve such problems.

What is a chess task? It is a chess problem where a person
seeks an objective in a way that maximizes or minimizes one or
more parameters. Among chess players the best-known task
question is: What is the shortest possible game? The answer, of
course, is the "fool's mate." White opens with, say, P–KB4.
Black replies P–K3. If White foolishly moves P–KN4, Black
checkmates on his second move, Q–R5.

The shortest game ending in perpetual check was published
in 1866 by one of the great pioneer chess problemists, Sam
Loyd. It is

1.	P–KB4	1.	P–K4
2.	K–B2	2.	Q–KB3
3.	K–N3	3.	Q×P (ch)

Black now has a perpetual check by moving his queen back and forth from the square it is on to Black's R3 square.

A much more difficult task was also posed in 1866 by Loyd. What is the shortest game ending in stalemate? Loyd's spectacular 10-move solution has never been surpassed:

White	*Black*
1. P–K3	1. P–QR4
2. Q–R5	2. R–R3
3. Q×QRP	3. P–KR4
4. Q×BP	4. QR–KR3
5. P–KR4	5. P–KB3
6. Q×QP (ch)	6. K–B2
7. Q×NP	7. Q–Q6
8. Q×N	8. Q–KR2
9. Q×B	9. K–N3
10. Q–K6 (stalemate)	

The final position is shown in Figure 116, No. 1. In 1882 a search began for the shortest "no capture" stalemate that left all 32 men on the board. The present record, 12 moves, was found by C. H. Wheeler in 1887. It was forgotten, then rediscovered independently by several men, including Loyd and Henry Ernest Dudeney (who gives it as Problem 349 in his *Amusements in Mathematics*). In January, 1906, Loyd published in *Lasker's Chess Magazine* a hilarious commentary on the game, pretending to explain the strategy behind each crazy move and calling attention to a five-move mate overlooked by Black when he made his final stalemating move. (Loyd's commentary can be found in Alain C. White's *Sam Loyd and His Chess Problems*, 1913, pages 128–129, currently available as a Dover reprint.)

Figure 116, No. 2 shows how 30 men, the largest number known, can be placed in a legal position—a position that can result in actual play—such that no move is possible by either side: a double stalemate. It was published in 1882 by G. R. Reichelm, who also showed how the position could be reached in 25 moves. Note the pattern's twofold symmetry.

Another remarkable task solved by Loyd is to play the shortest game ending with only the two kings on the board. Loyd's 17-move solution is given in Alain White's book as Problem 116. The two kings are left on their own pawn squares. Different 17-move solutions were later found by others, with the

Figure 116, No. 1 & 2

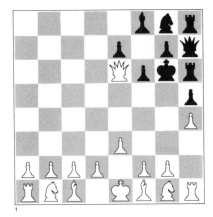

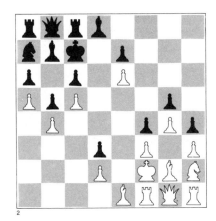

1. Shortest stalemate game   2. Double stalemate with 30 men

kings left on other cells. No one has found a 17-move game leaving the kings on their own starting squares. The two-king ending is rare among task problems in that 17 moves (by each player) can be proved an absolute minimum. Fifteen captures must be made by each side, but neither player can capture on his first move, and one more noncapture move can be proved necessary.

Dudeney later found a 17-move game (Problem 352 of his *Amusements in Mathematics*) that eliminates only the 14 pieces (nonpawns) of both sides, leaving both kings and the 16 pawns on their starting cells. Curiously, every move by Black is a mirror copy of White's preceding move. Here again, 17 moves can be proved minimal.

Along similar lines, one of Dudeney's great achievements was a 16-move game ending with all 16 of White's men on their starting cells and Black with only his king on the board. After Dudeney published this game [*see Figure* 116, *No.* 3] Loyd discovered that White could checkmate in three moves. This is another minimum, since no shorter mate is believed possible with Black's lone king on any other cell. Can the reader work out the mate before it is revealed in the Answer Section? Dudeney's game (Problem 351 of his *Amusements in Mathematics*) was reduced by a half-move in 1898—that is, the final position is achieved after White's 16th move—but then there is no mate in three because it is Black's turn.

A special class of task problem is known as a "one-move construction task" because only immediately possible moves are

Figure 116, No. 3 & 4

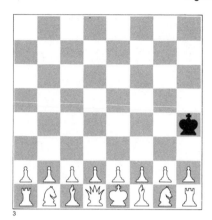

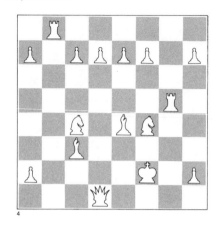

3. White to mate in three moves   4. 122 moves

considered. A classic example is the task of placing the eight
pieces of one color so that the largest number of moves can be
made. The proved maximum of 100 was achieved by M. Bezzel
in 1848 (see page 62 of *The Sixth Book of Mathematical Games
from Scientific American*). If all 16 men of one color are used, the
maximum was believed for 10 years to be 119 moves until
Nenad Petrovic increased it in 1949 to 122 [*see Figure* 116, *No.*
4]. When I first saw this pattern, I was unable to count more
than 104 moves until I realized that a promoted pawn must become one of four different pieces, each of course a different
move. (Modern chess laws do not allow a pawn on the eighth
rank to remain a pawn.) The record for the 16 black and white
pieces is 173, for all 32 men it is 164, and for a legal position
with no promoted men or promotion moves it is 181. The
present record for an illegal position is shown in Figure 116,
No. 5. By arranging the colors of the border queens as shown,
W. A. Shinkman, in 1923, achieved 412 moves. Captures are,
of course, counted as moves.

The minimum number of moves for the eight pieces of one
color is 10 (see my *The Unexpected Hanging and Other Mathematical Diversions*, page 88). The same position also minimizes the
number of pieces (three) among the eight that are able to
move. Ten also is the record for minimum moves when the 16
pieces of both colors are used. In 1923 T. R. Dawson found
the record minimum for all 32 men in a *legal* position [*see Figure* 116, *No.* 6]. Only two moves can be made. E. Fielder
showed in 1938 how the same 32 men can be legally placed so

Figure 116, No. 5 & 6

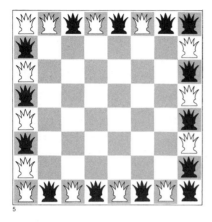

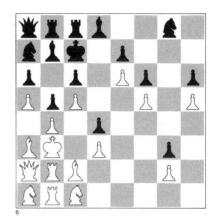

5. 412 moves   6. Two moves

that only one man (the white queen) can move [*see Figure* 116, *No.* 7]. No one has yet found a way to place legally all 32 men so that no move is possible.

There are many legal ways to place the 16 nonpawns to achieve a maximum of 46 captures, and all 32 men can be legally placed to allow 88 captures. How about illegal positions? If 32 black knights go on black cells and 32 white knights on white cells, 336 captures are possible. This was considered the

Figure 116, No. 7

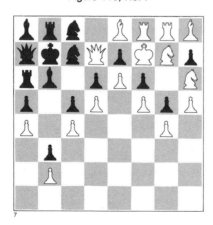

7. Only white queen can move

maximum for many decades until 1967, when T. Marlow ingeniously substituted two queens and two pawns for four
knights to raise the record to 338 [*see Figure* 116, *No.* 8]. It is
assumed that each capture by a pawn counts as four moves because it can become any of four pieces.

I have touched on only a small fraction of tasks concerning
moves and captures. Space does not allow discussing the
hundreds of tasks involving checks, discovered checks, mates,
selfmates, stalemates, forced captures (every move a capture),
forced checks, forced mates and so on. Of special interest to
combinatorial mathematicians are one-move construction tasks
involving the placing of a specified set of men so that a maximum or a minimum number of cells are attacked or unattacked, or to achieve some other goal that does not involve
moves or captures. A classic problem of this type (see Chapter
16 of *The Unexpected Hanging*) is to place eight queens (the
maximum) so that no queen attacks another. The similar tasks
of maximizing the number of nonattacking rooks (8), bishops
(14), knights (32) and kings (16) are considered in the same
chapter. A more difficult problem is to place 16 pawns (the
maximum) so that no three are in a straight line. Lines are not
restricted to rows, columns or diagonals but may have any orientation. Think of each pawn as a point in the center of the
cell it occupies. No three such points may be colinear. One of
many solutions is shown in Figure 116, No. *9*. It is the only one
in which two pawns occupy central cells.

Figure 116, No. 8 & 9

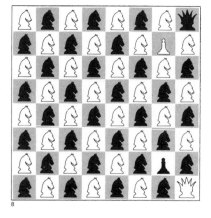

8. 338 captures   9. No three pawns in a line

Another difficult task of the same general category is to place eight queens so that 11 vacant squares are not attacked. There are at least six basic ways to do it (the exact number is not known), one of which will be given in the Answer Section. Eleven unchecked cells is undoubtedly maximum, although no proof is known to me.

A generalization of this problem—placing n queens on an order-n square to leave a maximum number of unattacked vacant cells—has not, to my knowledge, been fully analyzed. When n equals 1, 2 or 3, it is easy to see that no cell may be unchecked. When n equals 4, only one cell may be unchecked. For n equals 5 the problem is suddenly nontrivial. Three cells may be unattacked, but the pattern is difficult to find and also unique, except for rotations and reflections. Can the reader find it before checking the answer? The maximum number of unattacked cells when n equals 6, 7, 8, 9, 10, 11, 12 is believed to be 5, 7, 11, 16, 22, 27 and 36 respectively.

The minimum number of queens needed to attack all vacant cells of square boards is a general problem that has been thoroughly explored for boards of order 2 through 13. Since no piece attacks the cell it is on, the problem falls into three main groups: solutions in which no queen attacks another, or all queens are attacked, or some, but not all, are attacked. On the standard chessboard five queens are required in all three cases, and there are hundreds of solutions. Two tasks of this type on smaller boards are particularly pretty because each has only one basic solution. Can the reader put three queens on an order-6 board so that all vacant cells are attacked? Can he put four queens on an order-7 board so that all vacant cells are attacked and no queen attacks another?

Four queens can be placed on the order-8 board so that a maximum of 58 vacant cells are checked, leaving only two unchecked empty cells. There are many ways to eliminate those two squares by adding a single rook, bishop or king, but to check all vacant cells with four queens and a knight seems to have only one basic solution, which was first published by J. Wallis in 1908. Can the reader discover it? (Hint: The four queens must leave three of the vacant cells unchecked.)

It is easy to prove that nine kings, eight bishops or eight rooks are needed to attack all vacant cells on a standard chessboard. Much harder to find is the unique pattern by which 12 knights (the minimum) check all vacant cells. (See my *Mathematical Magic Show*, Chapter 14.) To attack all 64 squares requires 14 knights, or eight rooks, or 10 bishops, or 12 kings.

The eight pieces of one color can attack all 64 squares only if the bishops are on the same color. With bishops on opposite colors 63 squares is maximum. Douglas G. Smith, of Fresno, CA., recently sent me the result of his long search for a way to eliminate one of these eight pieces and still attack all vacant cells. He found how to do it by dropping a bishop. I do not know if his beautiful solution is unique (aside from rotations, reflections and trivial rearrangements of the rooks and queen) or if the task can be solved by dropping a knight or the king instead of a bishop. To make the task completely clear: Place a queen, king, two rooks, two knights and a bishop on a chessboard so that all vacant cells are in check.

For readers who may want to go more deeply into this obscure corner of chess recreations, I have listed basic references in the Bibliography for this chapter.

ANSWERS

1. Sam Loyd's three-move mate, all white men in starting position and a lone black king on Black's KR5:

1. P–Q4	1. K–R4
2. Q–Q3	2. K moves
3. Q–KR3 (mate)	

or

1. P–Q4	1. K–N5
2. P–K4 (ch)	2. K moves
3. P–KN3 (mate)	

2. One of six known ways to place eight queens so that 11 vacant cells are unattacked is shown in Figure 117 *a*. The unchecked squares are indicated by dots.

3. There is only one basic way to place five queens on an order-5 board so that three vacant cells are unattacked [*see Figure* 117 *b*]. Mannis Charosh has suggested that the best systematic search procedure for proving uniqueness is to explore the equivalent problem of placing three queens so that five cells are unchecked, taking advantage of the board's symmetries to shorten the search.

4. The only basic way to place three queens on an order-6 board so that all vacant cells are checked is shown in Figure 117 *c*.

5. The only basic way to put four queens on an order-7 board so that all vacant cells are checked and no queen attacks another queen is shown in Figure 117 *d*.

Figure 117

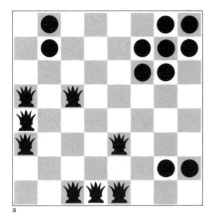

a

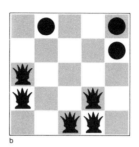

b

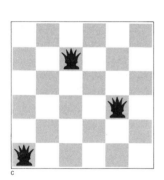

c

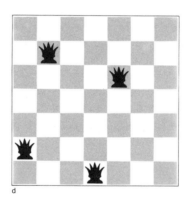

d

e

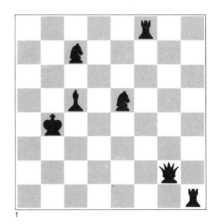

f

Answers to the chess tasks

6. The only known basic way to place four queens and one knight so that all vacant cells are attacked is shown in Figure 117 *e*.

7. Figure 117 *f* shows one way to place seven of the eight pieces of one color so that all vacant cells are in check. The positions of the rooks and queen can be given trivial variations.

ADDENDUM

The problem of placing five queens on a 5 x 5 board so that three cells are not attacked has appeared in many places since I introduced it in my 1972 column. It is usually given in the following form: Place five queens of one color and three of another color on an order-5 board so that no queen attacks a queen of a different color. I myself gave it in this form in a later (February 1978) column.

I had in 1972 confined the task to n queens on a board also of order n. When I gave it again in 1978 for the order-5 board, a number of readers generalized it to k queens on an order-n board. The best results came from Hiroshi Okuno, of Tokyo, whose computer search provided valuable data for low values of n and k. In 1983, Ronald L. Graham and Fan K. Chung, both of Bell Laboratories, turned their attention to Okuno's data and made some truly remarkable discoveries about the general problem. They will be reported in a paper that may appear before this book is off the press.

BIBLIOGRAPHY

"Chessboard Recreations." W. W. Rouse Ball in *Mathematical Recreations and Essays*. Macmillan, 1892, twelfth edition, 1972.

Les tours de force dur l'échiquier. Alain White. Paris: 1906.

"Chessboard Problems." Henry E. Dudeney in *Amusements in Mathematics*. London: Thomas Nelson, 1917, Dover reprint, 1958.

Ultimate Themes. Thomas Rayner Dawson, privately published by C. M. Fox, Surrey, England, 1937; included in *Five Classics of Fairy Chess*, by Dawson, Dover, 1973.

Rund um das Schachbrett. Karl Fabel. Berlin: Walter De Gruyter, 1955.

"Combinatorial Problems on the Chessboard." A. M. Yaglom and I. M. Yaglom, in *Challenging Mathematical Problems*. Holden-Day, 1964.

"Chessboard Placement Problems." Joseph Madachy in *Mathematics on Vacation*. Scribner's, 1966.

Schach und Zahl. Eero Bonsdorff, Karl Fabel, and Olavi Riihimaa. Düsseldorf: Walter Rau Verlag, 1966.

A Guide to Fairy Chess. Anthony S. M. Dickins. Privately printed by the author, Surrey, England, 1967. Revised edition, Dover, 1971.

Records in One-Mover Chess Construction Tasks. W. Cross and A. S. M. Dickins. Privately printed, Surrey, England, 1970.

"Chess Pieces." Stephen Ainley in *Mathematical Puzzles.* London: G. Bell, 1977.

Chess Ultimates. An irregularly published periodical on chess tasks, edited by Thur Row (Morton W. Luebbert, Jr.), 12039 Gardengate Drive, St. Louis, Mo. 63141. The cost is 50¢ per copy in the United States and Canada, 65¢ elsewhere. A *Chess Ultimates Handbook* with more than 1,000 record positions is in preparation.

18

SLITHER, 3X+1

AND OTHER CURIOUS QUESTIONS

> Pride in craftsmanship obligates the
> mathematicians of one generation to dispose
> of the unfinished business of their
> predecessors.
>
> —E. T. BELL, *The Last Problem*

Two familiar irrational numbers are π (3.141 . . .), the ratio of
the circumference of a circle to its diameter, and e (2.718 . . .),
the base of natural logarithms. Each has a nonrepeating decimal fraction. Both π and e are also transcendental numbers,
that is, numbers that are not algebraic. Specifically, a transcendental number is an irrational number that is not the root of
an algebraic equation with rational coefficients. Is the sum of π
and e transcendental? No mathematician knows if the sum is
even irrational.

One might suppose that any two numbers with infinite, nonrepeating decimal fractions would necessarily have a sum with
a nonrepeating (therefore irrational) decimal fraction. This is
not the case. The difference between π and 7, for instance, is
another transcendental. It is easy to compute. Represent 7 as
6.999 . . . , then subtract π (3.14159 . . .) to obtain the transcendental number 3.858407. . . . The sum of these two transcen-
dentals obviously is 6.999 . . . , or 7.

It seems unlikely, but until someone proves otherwise π and
e could be related by a curious unknown formula that would
give their sum a repeating (rational) decimal fraction with a
very long period, say one of more than a billion digits. It also
is not known if πe, π^{π}, e^{e} or π^{e} are irrational. It *has* been shown,

however, that e^π is transcendental, and it is easy to prove that at least one of the two numbers, πe and $(\pi + e)$, is transcendental. The unanswered questions about π and e are among hundreds of problems that are ridiculously simple to state but so difficult and deep that long-lasting fame awaits the first person to solve them.

It is not easy to distinguish significant unsolved problems from trivial ones. In *A Mathematician's Apology,* G. H. Hardy characterized a significant problem as being one connected to such a large complex of other mathematical ideas that when it is solved, it leads to important advances in mathematics and perhaps in science as well. An example of an essentially trivial but extremely difficult question is: If two people play the best possible checker game, will it end in a draw, a victory for the player who makes the first move or a victory for the player who makes the second move? A computer, given enough time, will probably work out the answer one day. When it does, the solution is unlikely to lead to any breakthroughs in mathematics or science. On the other hand, settling Fermat's last theorem would open all kinds of barred doors. (Please do not send me proofs. I am incapable of spotting flaws and always return them unread.)

There are dozens of unsolved map-coloring problems that, although they may not be as profound as the recently solved four-color theorem, are by no means trivial. Here is a notorious one given in C. Stanley Ogilvy's new revision of *Tomorrow's Math,* a splendid collection of unsolved problems for amateurs. What is the minimum number of colors needed to color the plane in such a way that any pair of points a unit distance apart are in regions of different colors? The question was first raised 20 years ago by Paul Erdös, a prolific inventor of problems.

That such a map must have at least four colors was cleverly established by Leo Moser with the diagram in Figure 118. Each edge of this graph has a length of one unit. Imagine that the graph is placed anywhere on a plane in which the problem is solved with only three colors. If vertex a is on red, say, then b and c must be on the other two colors, and g also must be red. Similarly, d and e must be on the other two colors and f must be red. Now, however, we have contradicted our assumption, because f and g, which are a unit distance apart, are both on red. At least four colors are therefore necessary.

Ogilvy's book has a proof that seven different colors are enough [*see Figure* 119]. The numbers give a repeating color pattern for a hexagonal tesselation of the plane, each hexagon

Figure 118

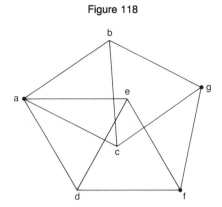

Leo Moser's graph for proving four-color necessity

Figure 119

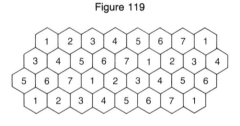

Proof of seven-color sufficiency

slightly less than one unit from corner to opposite corner. The gap remaining to be closed is a big one. Do such maps of four, five or six colors exist? No one yet knows.

There is an unusual class of unsolved arithmetic problems that, to use a computer-science term, we can call "looping" problems. A series of integers is generated according to a rule. One then asks if the series always enters one or more loops in which a finite set of integers keeps repeating cyclically. For example, start with any positive integer. Halve it if it is even; triple it and add 1 if it is odd. Keep repeating this procedure until the series loops in the cycle 2, 1, 4, 2, 1, 4, ... (Sample: 3, 10, 5, 16, 8, 4, 2, 1, ...) But does the series always enter the 2, 1, 4 loop? No one has proved that it does, nor has a counterexample been found.

Since Ogilvy revised his book a group of workers in the Artificial Intelligence Laboratory at the Massachusetts Institute of Technology have computer-tested all positive integers to 60,000,000 without finding an exception. They also discovered that if the rule $3n + 1$ (for odd integers) is replaced by $3n - 1$, the result, in absolute values, is the same as starting with a negative integer and following the old rules. In this case all negative integers to $-100,000,000$ were found to enter one of three loops with the following absolute values:

1. 2, 1, 2, 1, ...
2. 5, 14, 7, 20, 10, 5, ...
3. 17, 50, 25, 74, 37, 110, 55, 164,
   82, 41, 122, 61, 182, 91, 272,
   136, 68, 34, 17, ...

Michael Beeler, William Gosper and Rich Schroeppel give these results in HAKMEM (short for "Hacker's Memo"), Memo 239, Artificial Intelligence Laboratory, M.I.T., 1972, page 64. No one has yet come up with good ideas about how to establish the general case for all nonzero integers. (Zero, of course, is already in a 0, 0, 0, . . . loop.) No one knows if there are other loops, or if there are integers that generate a nonlooping series of numbers that diverge to infinity.

Gosper and Schroeppel, incidentally, proved an amusing loop conjecture involving English names for numbers (HAK-MEM, page 64). Spell out the name of any number. It need not be rational or even real. Counting numbers must be named directly, and not by such circumlocutions as "twelve plus one" or "twenty minus five," and so on. Replace the name by the number of digits in the name and keep repeating the procedure. Example: THE CUBE ROOT OF PI, FIFTEEN, SEVEN, FIVE, FOUR, FOUR, FOUR, ... The series always, and quickly loops at FOUR.

In explaining a recent triangle-dissection problem, Ogilvy wrote that it might "have a solution before this book appears." He was right. It had been known that any triangle can be cut into four triangles similar to itself, or into n triangles similar to itself when n is 6 or more. If n is 2 or 3, only a right triangle can be properly cut. If n is 5, a right triangle can be dissected into five triangles similar to itself, but for nonright triangles the conditions for dissection were unknown when R. W. Freese, Ann K. Miller and Zalman Usiskin wrote their article "Can Every Triangle be Divided into n Triangles Similar to It?" (*The American Mathematical Monthly*, Volume 77, October, 1970, pages 867–869).

Recently it has been independently proved by several people that when n is 5 and the triangle has no right angle, it can be cut into five triangles similar to itself if and only if one angle is 120 degrees and the others are each 30 degrees [*see Figure* 120]. This unique dissection is given in "Partitioning a Trian-

Figure 120

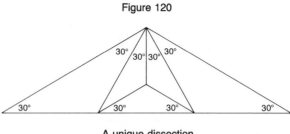

A unique dissection

gle into 5 Triangles Similar to It" (*Mathematics Magazine*, Volume 45, January, 1972, pages 37–42), by Z. Usiskin and S. G. Wayment. Still open are questions such as: Which triangles can be cut into n similar triangles *not* similar to themselves? For what values of n can a quadrilateral be cut into n quadrilaterals similar to one another and/or to itself?

In 1960, Stanislaw M. Ulam, another virtuoso puzzle maker, published a fine collection of advanced unsolved problems, most of them original. The book was reprinted in 1964 as a paperback, *Problems in Modern Mathematics*. One of Ulam's topological-game problems seemed as uncrackable as it was curious. Imagine a cube divided into a lattice of unit cubes, like a three-dimensional checkerboard. Players take turns marking a unit edge of the lattice. The first player marks any edge. Thereafter each marked edge must join the previously marked edge. One end of the path remains fixed as the other end grows one unit in length with each move, as though a bug were crawling along the lattice lines and leaving a trail. Since the lattice is finite, the path must eventually intersect itself to form a closed-space curve. One of the players wins if the curve is knotted. The other player wins if there is no knot. Who wins when the game is played rationally?

John Horton Conway, a University of Cambridge mathematician, found an ingenious proof that the "no knot" player can always win regardless of whether he goes first or second. Assume that the game is played on a three-by-three-by-three cube (a lattice of 27 unit cubes). This is the smallest cube on which the path can knot. The following no-knot strategy extends readily to all cubes of higher orders.

Through each lattice point there are various planes that are perpendicular to a body diagonal (a diagonal joining diametrically opposite corners) of the large cube. We shall call such a plane a primary plane, or P plane. If the P plane goes through a corner of the large cube, there may be only one adjacent plane parallel to it and passing through points adjacent to points on the P plane; otherwise there will be two such adjacent planes, one on each side of the P plane. We call these A planes.

Imagine all lattice points on A planes—call them A points—projected on the P plane, together with all edges joining A points to P points. This puts on the P plane a graph equivalent to one of the five shown in Figure 121. On each graph black vertexes are lattice points originally on the P plane. The open circle vertexes are A points projected from A planes. The C's mark the corners of the large cube.

Figure 121

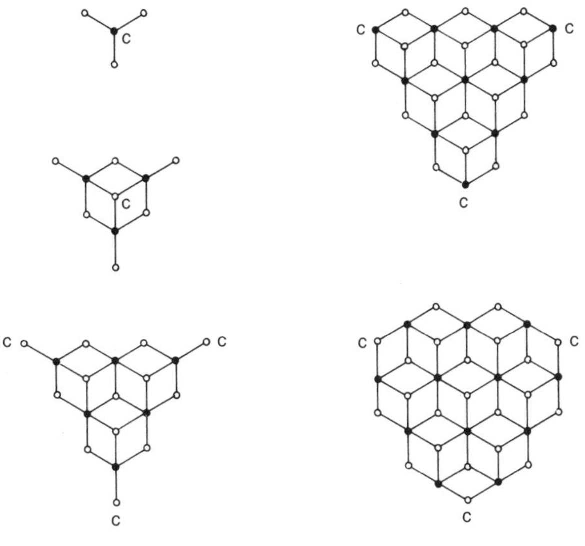

Graphs for Stanislaw M. Ulam's knot game

Note that the three graphs on the left have loose ends. Those on the right do not. Conway has shown (his unpublished proof is not difficult) that, given any lattice point, one can always find passing through it a P plane on which the graph has no loose ends.

If your opponent goes first, you should choose a P plane through either end of the marked edge that has a no-loose-ends graph. One end of the path will be on an A plane. Play so that the path returns to the P plane, that is, join the black end of the path to a black vertex. Each succeeding move by your opponent must take the path off the P plane (to a black spot).

Your strategy is always to return the path to the P plane by extending it to a black spot. Because the graph has no loose ends, and because its black and open-circle vertexes alternate, you can always do so. It is obvious that when the path first closes, it will have to be unknotted.

If you go first, mark any edge. After your opponent has moved choose a P plane that goes through the path's middle vertex and on which the graph has no loose ends. Your no-knot strategy is the same as before. Play so that the path always returns to the P plane; in other words, always extend the path to a black vertex. The path cannot be knotted when it first intersects itself.

"I think it was obvious from the start," Conway writes in a letter, "that the no-knot player had the best of it. *He* only had to make the path close, whereas the other player really had to *do things*."

Conway's strategy does not apply to noncubical "brick" lattices (because finding a no-loose-ends graph is not always possible) or to cubical games in which the no-knot player goes first and moves are allowed at all times at either end of the growing path. In both cases, so far as I know, winning strategies remain unknown.

Three-dimensional lattices are awkward "boards" for actual play, but closely related topological games on planar lattices make excellent pencil-and-paper contests. David L. Silverman, whose book *Your Move* includes several such games, is responsible for the latest fad among Los Angeles puzzlers: an unpublished, unsolved game that Silverman calls Slither. Its five-by-six-point lattice is just large enough to have resisted all efforts to determine which player has the win [*see Figure* 122]. In a tab-

Figure 122

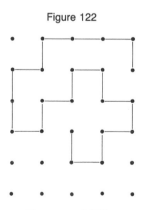

The game of Slither

ulation of several hundred games the wins were about equally divided between first and second players. The rules are simple. Opponents take turns marking an orthogonal unit segment. The segments must form a continuous path but may be added to either end of the preceding path. The player forced to close the path is the loser. (If the first to close it wins, it is a duller game, although even *that* version is unsolved.) The illustration shows a typical position in which the next play must be a losing one. Perhaps a reader will discover a winning strategy for either version (or both versions) of Slither.

Hallard T. Croft, a colleague of Conway's at Cambridge, periodically sends lists of new unsolved problems to his friends. A few years ago one of Croft's problems asked if there existed a finite set of points on the plane such that the perpendicular bisector of the line segment joining any two points would always pass through at least two other points of the set. The problem was solved by Leroy M. Kelly, a mathematician at Michigan State University. Although the problem cannot be called significant, Kelly's solution, using only eight points, is so elegant that I give it as an exercise.

ANSWERS

The solution for the problem of placing eight points so that the perpendicular bisector of each pair of points passes through at least two other points is shown in Figure 123.

David L. Silverman's game of Slither produced a flood of strategies of steadily mounting generality until finally Ronald C. Read, a graph theorist at the University of Waterloo, reduced the standard game to monumental triviality.

Figure 123

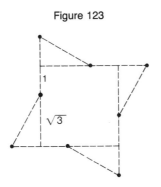

Solution to the eight point problem

Standard Slither is played on a rectangular field consisting of a square lattice of dots. Players take turns drawing orthogonal unit "edges" connecting adjacent dot pairs, adding each move to either end of the continuous path that is formed. The player who is forced to make the slithering line meet itself is the loser. Several dozen readers immediately pointed out that on the 5-by-6 field that had been given as a sample playing field, the first player has an easy win by taking the central edge and thereafter making his moves symmetrically opposite to his opponent's moves. He also wins the reverse version (the first to close the path wins) by seizing the first winning opportunity offered.

George A. Miller of Philadelphia was the first to provide a general strategy for all rectangular boards. If the field has an even number of dots, draw a Hamiltonian path along the lattice lines, that is, a path visiting every dot once only. Color the alternate edges red, beginning and ending with red. The first player's winning strategy is: Always play red. If the field contains an odd number of dots, the second player's winning strategy is: After the first move, draw any Hamiltonian path starting at one end of the first move, color the path as before and always play red. Essentially the same strategy was also discovered by Michael Kelly, by Oliver G. Selfridge, and others.

Next I found that this strategy applies if diagonal moves between adjacent dots are allowed, and also when the game is played on triangular lattices. My elation was short-lived. When I wrote to Read about it, he saw at once that these were merely special cases of a general strategy that applies to any set of dots in any formation in a space of any dimensions. Moreover, a "move" can be the joining of any pair of dots, and it does not matter whether this is allowed at both ends of the path or only at the end of the preceding move.

Read explained it this way. A graph is said to have a "1-factor" if it is possible to join all the nodes in pairs so that every node belongs to one and only one of the disjoint edges. Think of an array of dots as the vertexes of a complete graph consisting of all possible joining edges. Draw a 1-factor of the graph with a red pencil. (A Hamiltonian path on square lattices is one way of doing this, but the 1-factor is more general because some graphs have 1-factors but no Hamiltonian path.) Any move connecting two dots is now allowed.

If the number of dots is even, the graph can be 1-factored and the first player wins by always playing on red edges. If the number of dots is odd, the second player disregards one end of the first move, 1-factors the remaining dots and plays always

on red. The player who first runs out of unused dots to move
to is the loser.

This obvious and trivial parity strategy was obscured in
Slither by the game's many irrelevancies. Reverse Slither, in
which the first to close the path *wins*, is a more difficult matter.
As we have seen, a symmetry strategy wins for the first player
on all odd-by-even fields. The second player can win by bilateral symmetry play if the first play is to a main diagonal of a
square or to a central orthogonal line of any rectangle that has
one. Selfridge has found a strategy for a second-player win on
all squares.

Michael Beeler of M.I.T. wrote a computer program for reverse Slither. Here are some of its results:

1. The second player wins on squares through order 6.

2. Taking the center move is the only winning first play on
the 3 by 4, 4 by 5, 4 by 9 and 5 by 6.

3. A theory devised by Beeler, establishing a first-player win
on all $2 \times n$ fields (n greater than 2) is confirmed through
$n = 18$.

4. On $3 \times n$ fields, $n = 2$ through 12, the first player wins if n
is even, loses if n is odd.

5. On $4 \times n$ fields, $n = 5$ through 9, the first player wins in all
cases.

6. The second player wins on the 5 by 7.

These data suggest the following unproved conjecture: The
first player wins on all nonsquare rectangles if the number of
spots is even. The second player wins on all squares and on all
nonsquare rectangles if the number of spots is odd.

ADDENDUM

The $3X + 1$ problem, as it is now usually called, is still resisting
solution. According to Richard Guy, it was first proposed before World War II by Lothar Collatz, now a mathematician at
the University of Hamburg, when he was a student. In a 1970
lecture H. S. M. Coxeter offered $50 for a proof he could un-
derstand or $100 for a counterexample. He has since been del-
uged with so many false proofs that he is no longer willing to
evaluate them. Indeed, it seems as easy to make subtle mistakes
in such proofs as in proofs of Fermat's last theorem. False
proofs have even been published; for example, in *Fibonacci
Quarterly*, Vol. 18, 1980, pages 231–242. In 1982 Paul Erdös
expressed his opinion—and who is more qualified to give

one?—that if the conjecture is true, present-day number theory lacks the tools for a proof.

A counterexample would be a number that either keeps generating larger and larger numbers forever, without ever repeating a number, or one that enters a loop higher than the 4-2-1. If a counterexample exists it would have to be exceedingly large because the conjecture has been tested, according to Guy, for all numbers less than 7×10^{11}. Early in the game it was observed that it is not necessary to test even numbers, or odd numbers of the form $4k+1$, $16k+3$, or $128k+7$. This greatly simplifies computer programs. Of course as soon as a sequence hits a power of 2, often after many chaotic ups and downs, it crunches quickly to 4-2-1. The power of 2 on which most sequences converge is 16.

Among numbers smaller than 50, the worst is 27. After 77 steps it reaches its peak of 9,232, then 34 steps are required to take it down to 1. When John H. Conway introduces the $3X+1$ conjecture in lectures, he likes to stand by a blackboard and say, "Let's take some random small number, say 27, and see what happens." A graph theorist would describe the theorem by saying that, if true, we can draw an infinite directed tree, each point labeled with a distinct positive integer, that will catch all the integers, and which will converge along the arrows to a root that is the triangular cycle 4-2-1.

A simple proof that of the two numbers πe and $\pi + e$, at least one is transcendental, is given by David Brubaker in *Mathematics Magazine,* Vol. 44, November 1971, page 267.

On triangles that can be cut into five similar triangles *not* similar to the original, see Guy's report in *The American Mathematical Monthly,* Vol. 80, December 1973, page 1123. Apparently there are ten essentially different cases. The equilateral triangle and any isosceles triangle can be cut into five similar right triangles, and the equilateral triangle can also be cut into five similar triangles containing an angle of 120 degrees.

For more on "honest numbers" that spell with the same number of letters as the number they represent (FOUR is the only honest number in English), see Chapter 7 of my *Incredible Dr. Matrix.*

An ultimate generalization of Slither was analyzed by William N. Anderson, Jr., in a paper listed in the Bibliography. The game is played on an arbitrary finite graph, each player taking an edge of the graph on his turn. Anderson presents a strategy for this generalized Slither that is based on a matching algorithm in a 1965 paper by J. Edmonds.

BIBLIOGRAPHY

Problems in Modern Mathematics. Stanislaw M. Ulam. John Wiley, 1960, revised edition, 1964.

Tomorrow's Math. C. Stanley Ogilvy. Oxford University Press, 1962, revised edition, 1972.

Unsolved Problems in Number Theory. Richard K. Guy. Springer-Verlag, 1981.

On the $3X + 1$ problem:

Modern Abstract Algebra. Richard V. Andree. Holt, Rinehart and Winston, second edition, 1971.

Computer Approaches to Mathematical Problems. Jürg Nievergelt, J. Craig Farmer, and Edward M. Reingold. Prentice-Hall, 1974, pages 211–217.

"Problem 133." Comments by Charles Trigg and others. *Crux Mathematicorum,* Vol. 2, 1976, pages 144–150.

"Crazy Roller Coaster." William J. Bruce. *The Mathematics Teacher,* Vol. 71, January 1978, pages 45–49.

"On the $3X + 1$ Problem." R. E. Crandall. *Mathematics of Computation,* Vol. 32, 1978, pages 1281–1292.

Gödel, Escher, Bach. Douglas Hofstadter. Basic Books, 1979, pages 400–402.

"$3X + 1$ Revisited." Fred Gruenberger. *Popular Computing,* Vol. 7, October 1979, pages 3–12. (Earlier references in the same periodical are cited.)

Unsolved Problems in Number Theory. Richard K. Guy. Springer-Verlag, 1981, pages 120–121.

"On the Collatz $3n + 1$ Algorithm." L. E. Garner. *Proceedings of the American Mathematical Society,* Vol. 82, 1981, pages 19–22.

"The $3X + 1$ Problem and Its Generalization." J. C. Lagarias, 1982, unpublished.

On Slither:

"Slither." Anonymous. *Function,* Vol. 1, April 1977, page 13; October 1977, pages 15–20.

"Maximum Matching and the Game of Slither." William N. Anderson, Jr. *Journal of Combinatorial Theory,* Vol. 17(B), 1974, pages 234–239.

19

MATHEMATICAL TRICKS WITH CARDS

"Do you like card tricks?"
"No, I hate card tricks," I answered.
"Well, I'll just show you this one."
He showed me three.

—SOMERSET MAUGHAM, *Mr. Know-All*

Maugham's experience with card magicians is all too familiar. "I don't really like people who do card tricks," Elsa Maxwell once wrote (I quote from an autobiography of a lady magician, *You Don't Have to Be Crazy,* by Frances Ireland). "They never stop at one or two, but go on and on and on, and always make you take cards, or turn up cards, or cover cards, until you are worn out."

Mathematical card tricks, let it be admitted at once, are precisely the kind of tricks that are the most boring to most people. Nevertheless, they have a curious appeal to mathematicians and mathematically minded magicians.

Many excellent card deceptions are based on a parity principle, but the underlying even-odd structure is usually concealed so ingeniously that if you follow the directions with cards in hand you are likely to astonish yourself. Consider the following trick invented about 1946 by the Chicago card expert Ed Marlo. Magicians classify it as an "oil and water" effect, for reasons that will be apparent in a moment. There are many ways of achieving the same effect by secret and difficult "moves," but this version is entirely self-working.

Remove 10 red and 10 black cards from the deck and arrange them in two face-up piles, side by side, with all red cards. on the left and all black cards on the right. First you tell your watchers that you will demonstrate what you intend to do by using only five cards of each color. With both hands simultaneously remove the top card from each pile and place them, still face up, on the table at the bottom of each pile. Do the same with the next two top cards, but this time cross your arms before you place the two cards on the two new piles you are starting. This puts a black card on the red one and a red card on the black one. The next transfer of a pair of cards is made with uncrossed arms, the next with crossed arms, and the fifth and last pair is dealt with arms uncrossed. In other words, five simultaneous deals are made, with arms crossed only on alternate deals. On each side you now have a pile of five face-up cards with their colors alternating. Put either pile on the other one. Spread the 10 cards to show that colors alternate throughout.

Square the cards and turn the packet face down. From its top deal the cards singly and face up to form two piles again, dealing alternately to the left and right. Call attention to the fact that this procedure naturally separates the colors. At the finish you will have five reds on the left and five blacks on the right.

State that you will repeat this simple series of operations with all 20 cards. Begin as before, with 10 face-up reds on the left and 10 face-up blacks on the right. Transfer the cards to form two new piles, just as you did before, crossing your arms on alternate deals so that the colors alternate in each pile. After all 20 cards are dealt put one pile on the other, square the cards, turn the packet over and hold it face down in your left hand.

Deal 10 cards face up to form two piles, dealing from left to right and observing aloud that this brings the reds together on the left and the blacks together on the right. After the 10 cards have been dealt face up do not pause but continue smoothly and deal the remaining 10 cards face down. It is best to put down the cards so that they overlap in two vertical rows [*see Figure* 124].

Pick up the five face-down cards on the left with your left hand and the five face-down cards on the right with your right hand. Cross your arms and put the cards down. You explain that you have transferred half of the cards of each pile to the pile of the opposite color but that like oil and water the colors mysteriously refuse to mix. Turn over the face-down cards. To everyone's surprise (you hope) the reds are back with the reds and the blacks are back with the blacks! Readers should have little difficulty discovering why it works with any set of cards

Figure 124

The oil-and-water effect

containing an even number of cards of each color and why it did not work when you demonstrated it with 10 cards.

After you have finished the oil-and-water trick put the two piles together with either color on top. Turn the packet face down and spread it in a fan. You are ready to perform a red-black trick invented by Karl Fulves and published in his magic periodical, *The Pallbearers Review,* September, 1971.

Ask someone to pull slightly forward any 10 cards he pleases. The fan will resemble the one shown in Figure 125. With your right hand count the jogged (protruding) cards to make sure there are 10. Do this by removing the cards one at a time from right to left, putting them into a face-down pile as you count from one to 10. Close up the 10 cards remaining in your left hand and place them in a second face-down pile alongside the first.

Figure 125

Ten cards jogged forward

Tell your audience that an amazing thing has happened. Although 10 cards were selected randomly, the colors in the two piles are so ordered that every nth card in one pile has a color opposite to the color of the nth card in the other pile. To prove this, turn over the top cards of each pile simultaneously. One will be red and the other black. Place the black under the red, turn the pair over and put it aside to form a new face-down pile. Repeat the procedure with the cards now on top of the two original piles. They will be red-black too. Indeed, every pair you turn will be red-black!

As you show the pairs always put the black card under the red before you turn them over and place them on the third pile. When you finish, the cards in this face-down pile will have alternating colors.

Now you are ready to perform a truly mystifying trick in which parity is conserved in spite of repeated shuffling. Known as Color Scheme, it was invented by Oscar Weigle, an amateur magician who is now an editor at Grosset & Dunlap. It sold as a manuscript in magic stores in 1949.

Give the packet of 20 cards to someone and ask him to hold it under the table where neither he nor anyone else can see the cards. Tell him to mix the cards by the following procedure. (It is known as the Hummer shuffle, after Bob Hummer, the magician who first used it in tricks.) Turn over the top two cards (not one at a time but both together as if they were one card), place them on top and cut the packet. Your assistant is to keep repeating this procedure of turn two, cut, turn two, cut for as long as he wishes. The procedure will, of course, result in a packet containing an unknown number of randomly distributed reversed cards.

With the cards still held under the table, tell your assistant to do the following. Shift the top card to the bottom. Then turn over the next card, produce it from under the table and place it on the table. This procedure is repeated—card to bottom, reverse next card and deal—until 10 cards have been dealt to the table. It will be apparent that the cards have become mysteriously ordered. All the face-up cards are the same color and all the face-down ones are of the opposite color.

The second and climactic half of the trick, which Weigle confesses is a "bare-faced swindle," now unfolds. Your assistant is still holding 10 cards under the table. Ask him to shuffle them by separating them into two packets; then, keeping all the cards flat (no card must be allowed to turn over), weave the two packets into each other in a completely random way. You can demonstrate how to do this by using the 10 cards already dealt. After your assistant has executed the shuffle a few times, ask him to turn over the packet and shuffle the same way a few more times. If he likes, he can give the packet a final cut.

Now he continues with the dealing procedure he used before: card to bottom, next card reversed and dealt. (The final card is reversed and dealt.) In spite of the thorough mixing the result is exactly the same as before. All the face-up cards match the former face-up cards in color, and the same is true of all face-down cards.

One of the oldest themes in card magic is to produce in some startling fashion a card that has been randomly selected and replaced. Here is a simple method that exploits a binary sorting technique. Fulves published it in his periodical in November, 1970.

Take 16 cards from a shuffled deck and spread them face down on the table without mentioning how many cards you are using. A viewer selects a card, looks at it and places it on top of the deck. The remaining cards in the spread are squared and put on top of the deck above the chosen card. Ask him to cut off about half of the deck, give or take half a dozen cards. Actually he can take between 16 and 32 cards. He hands this packet to you.

Hold the packet in both hands. As your left thumb slides the cards one at a time to the right, move your right hand forward and back so that every other card, starting with the first one, is jogged forward. The resulting fan of cards will resemble the one in Figure 125 except that the jogged cards are not randomly distributed. Strip all the projecting cards from the fan and discard them. Square the remaining cards and repeat the procedure, jogging forward all the cards at odd positions, starting with the first card. Strip them out and discard. Continue in this way until one card is left. Before turning it over ask for the chosen card's name. It will be the card you hold.

A completely different method of locating a selected card can be found in several books on card magic. Turn your back and instruct someone to cut a shuffled deck into three approximately equal piles. He turns over any pile and then reassembles the deck by sandwiching the face-up pile between the other two, which remain face down. He is told to remember the top card of the face-up pile. With your back still turned, ask him to cut the deck several times, then give it one thorough riffle-shuffle. The shuffle will of course distribute the face-up cards randomly throughout the deck.

Turn around, reverse the pack and spread it in a row. Look for a long run of face-up cards, remembering that a cut may have split the run so that part of it is at each end of the spread. The first face-down card above the run is the chosen one. Slide it from the spread, have the card named and then turn it over.

Our last trick, based on a curious shuffling principle discovered by Fulves, is presented as a gambling proposition. All cards of one suit (the suit can be chosen by the victim) are removed from the deck. Assume that the discarded suit is diamonds. The remaining cards are arranged so that each triplet has three different suits in the same order. (Card values are ignored.) Again the victim may specify the ordering. Suppose he chooses spades, hearts and clubs. The 39-card deck is arranged from the top down so that the suits follow the sequence spades, hearts, clubs, spades, hearts, clubs and so on.

Place the deck face up in front of the victim. Ask him to cut it in two packets and riffle-shuffle them together. As he makes the cut, note the suit exposed on top of the *lower* half. We shall call this suit k. After the single shuffle the deck is turned face down. The cards are now taken from the top three cards at a time, and each triplet is checked to see if it contains two cards of the same suit.

It is hard to believe, but:

(1) If k is spades, no triplet will contain two spades.

(2) If k is hearts, no triplet will contain two clubs.

(3) If k is clubs, no triplet will contain two hearts.

This assumes, of course, a spades-hearts-clubs ordering. If the ordering is otherwise, the three rules must be modified accordingly; that is, spades must be changed to whatever suit is at the top of each triplet, and so on. Let m stand for the suit that you know cannot show twice in any triplet, and a and b for the suits that can.

Before dealing through the deck to inspect the triplets, make the following betting proposition. For every triplet containing a pair of m's you will pay the victim \$10. In return he must agree to pay you 10 cents for every pair of a's or b's. It seems like a good bet for the victim, but it is impossible for you to lose, and the swindle can be repeated as often as you please. Just arrange the cards again and allow the victim to make the single riffle-shuffle. Naturally you always promise to pay him for doublets of the suit that you know cannot show. The fact that this suit may vary from deal to deal makes the bet particularly mystifying.

As Fulves has observed, the triplets have other unexpected properties. Of the triplets containing pairs the a's and b's will alternate; after a pair of a's the next pair will be b's and vice versa. Pairs of one suit always include a top card of the triplet. Pairs of the other suit always include a bottom card.

No explanation of these tricks will be given. Readers will find it stimulating, however, to analyze each trick to see if they can comprehend exactly why it operates with such uncanny precision.

ADDENDUM

Peter T. Sarjeant extended Fulves' shuffling trick to the four suits of a full deck. Arrange the cards so that from top down the sequence is a repetition of clubs, diamonds, hearts, spades. As before, the deck is placed face up and cut about in half. Note the suit on the top of the bottom half. Call it k. The halves are then interlaced with a single riffle-shuffle.

When cards are taken four at a time from the top you will find the following true of each quadruplet:

(1) If k is clubs, there will be no pair of hearts and no pair of clubs.

(2) If k is diamonds, any suit may be paired.

(3) If k is hearts, there will be no pair of diamonds and no pair of spades.

(4) If k is spades, any suit may be paired.

Knowledge of these facts can, of course, be the basis of a variety of betting swindles.

Edward M. Cohen proposed the following variation of Fulves' trick involving a selected card that goes sixteenth from the top of the deck. He likes to begin by forming a square array of 16 cards, face down on the table. A spectator picks a row. Another person picks a column. The card at the intersection is turned face up and remembered. This card goes to the *bottom* of the deck. The remaining 15 cards are swept into a pile and the deck placed on top of them. The chosen card is now sixteenth from the bottom.

Anyone may now cut the deck about in half (it is only necessary that the lower portion contain more than 16 and less than 32 cards). The top half is discarded. Hand the lower half to someone with the request that he deal it into two face-up piles, alternating piles as he deals. The pile that gets the *last* card is discarded. The other pile is turned face down, and this procedure is repeated until only one card remains. It will be the chosen card.

Hundreds of more elaborate card tricks have been based on the binary principles that underlie this trick, but the one just described is as simple, effective, and as easy to perform as any.

BIBLIOGRAPHY

Scarne on Card Tricks. John Scarne. Crown, 1950

Mathematics, Magic and Mystery. Martin Gardner. Dover, 1956.

Mathematical Magic. William Simon. Scribner's, 1964.

Self-Working Card Tricks. Karl Fulves. Dover, 1976.

20

THE GAME OF LIFE, PART I

Most of the work of John Horton Conway, a distinguished mathematician at the University of Cambridge, has been in pure mathematics. For instance, in 1967 he discovered a new group—some call it "Conway's constellation"—that includes all but two of the then known sporadic groups. (They are called "sporadic" because they fail to fit any classification scheme.) It is a breakthrough that has had exciting repercussions in both group theory and number theory. It ties in closely with an earlier discovery by John Leech of an extremely dense packing of unit spheres in a space of 24 dimensions where each sphere touches 196,560 others. As Conway has remarked, "There is a lot of room up there."

In addition to such serious work Conway also enjoys recreational mathematics. Although he is highly productive in this field, he seldom publishes his discoveries. One exception was his paper on "Mrs. Perkins' Quilt," a dissection problem discussed in my *Mathematical Carnival*. Another was sprouts, a topological pencil-and-paper game invented by Conway and M. S. Paterson. It is also the topic of a chapter in the same book.

In this chapter we consider Conway's most famous brainchild, a fantastic solitaire pastime he calls "Life." Because of its analogies with the rise, fall and alterations of a society of living organisms, it belongs to a growing class of what are called "simulation games"—games that resemble real-life processes. To play Life without a computer you need a fairly large checkerboard and a plentiful supply of flat counters of two colors. (Small checkers or poker chips do nicely.) An Oriental "go" board can be used if you can find flat counters small enough to fit within its cells. (Go stones are awkward to use because they are not flat.) It is possible to work with pencil and graph paper but it is much easier, particularly for beginners, to use counters and a board.

The basic idea is to start with a simple configuration of counters (organisms), one to a cell, then observe how it changes as you apply Conway's "genetic laws" for births, deaths and survivals. Conway chose his rules carefully, after a long period of experimentation, to meet three desiderata:

(1) There should be no initial pattern for which there is a simple proof that the population can grow without limit.

(2) There should be initial patterns that *apparently* do grow without limit.

(3) There should be simple initial patterns that grow and change for a considerable period of time before coming to an end in three possible ways: Fading away completely (from overcrowding or from becoming too sparse), settling into a stable configuration that remains unchanged thereafter, or entering an oscillating phase in which they repeat an endless cycle of two or more periods.

In brief, the rules should be such as to make the behavior of the population both interesting and unpredictable.

Conway's genetic laws are delightfully simple. First note that each cell of the checkerboard (assumed to be an infinite plane) has eight neighboring cells, four adjacent orthogonally, four adjacent diagonally. The rules are:

(1) Survivals. Every counter with two or three neighboring counters survives for the next generation.

(2) Deaths. Each counter with four or more neighbors dies (is removed) from overpopulation. Every counter with one neighbor or none dies from isolation.

(3) Births. Each empty cell adjacent to exactly three neighbors—no more, no fewer—is a birth cell. A counter is placed on it at the next move.

It is important to understand that all births and deaths occur *simultaneously*. Together they constitute a single generation or, as we shall usually call it, a "tick" in the complete "life history" of the initial configuration. Conway recommends the following procedure for making the moves:

(1) Start with a pattern consisting of black counters.

(2) Locate all counters that will die. Identify them by putting a black counter on top of each.

(3) Locate all vacant cells where births will occur. Put a white counter on each birth cell.

(4) After the pattern has been checked and double-checked

to make sure no mistakes have been made, remove all the dead counters (piles of two) and replace all newborn white organisms with black counters.

You will now have the first generation in the life history of your initial pattern. The same procedure is repeated to produce subsequent generations. It should be clear why counters of two colors are needed. Because births and deaths occur simultaneously, newborn counters play no role in causing other deaths or births. It is essential, therefore, to be able to distinguish them from live counters of the previous generation while you check the pattern to be sure no errors have been made. Mistakes are very easy to make, particularly when first playing the game. After playing it for a while you will gradually make fewer mistakes, but even experienced players must exercise great care in checking every new generation before removing the dead counters and replacing newborn white counters with black.

You will find the population constantly undergoing unusual, sometimes beautiful and always unexpected change. In a few cases the society eventually dies out (all counters vanishing), although this may not happen until after a great many generations. Most starting patterns either reach stable figures—Conway calls them "still lifes"—that cannot change or patterns that oscillate forever. Patterns with no initial symmetry tend to become symmetrical. Once this happens the symmetry cannot be lost, although it may increase in richness.

Conway originally conjectured that no pattern can grow without limit. Put another way, any configuration with a finite number of counters cannot grow beyond a finite upper limit to the number of counters on the field. This is probably the deepest and most difficult question posed by the game. Conway offered a prize of $50 to the first person who could prove or disprove the conjecture before the end of 1970. One way to disprove it would be to discover patterns that keep adding counters to the field: A "gun" (a configuration that repeatedly shoots out moving objects such as the "glider," to be explained below) or a "puffer train" (a configuration that moves but leaves behind a trail of "smoke"). The results of the contest for Conway's prize are discussed in the next chapter.

Let us see what happens to a variety of simple patterns.

A single organism or any pair of counters, wherever placed, will obviously vanish on the first tick.

A beginning pattern of three counters also dies immediately unless at least one counter has two neighbors. Figure 126 shows the five connected triplets that do not fade on the first tick.

Figure 126

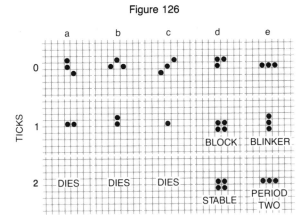

The fate of five triplets in "life"

(Their orientation is of course irrelevant.) The first three [a, b, c] vanish on the second tick. In connection with c it is worth noting that a single diagonal chain of counters, however long, loses its end counters on each tick until the chain finally disappears. The speed a chess king moves in any direction is called by Conway (for reasons to be made clear later) the "speed of light." We say, therefore, that a diagonal chain decays at each end with the speed of light.

Pattern d becomes a stable "block" (two-by-two square) on the second tick. Pattern e is the simplest of what are called "flip-flops" (oscillating figures of period 2). It alternates between horizontal and vertical rows of three. Conway calls it a "blinker."

Figure 127 shows the life histories of the five tetrominoes (four rookwise-connected counters). The square [a] is, as we have seen, a still-life figure. Tetrominoes b and c reach a stable figure, called a "beehive," on the second tick. Beehives are frequently produced patterns. Tetromino d becomes a beehive on the third tick. Tetromino e is the most interesting of the lot. After nine ticks it becomes four isolated blinkers, a flip-flop called "traffic lights." It too is a common configuration. Figure 128 shows 12 common forms of still life.

The reader may enjoy experimenting with the 12 pentominoes (all possible patterns of five rookwise-connected counters) to see what happens to each. He will find that five vanish before the fifth tick, two quickly reach a stable loaf, and four in

Figure 127

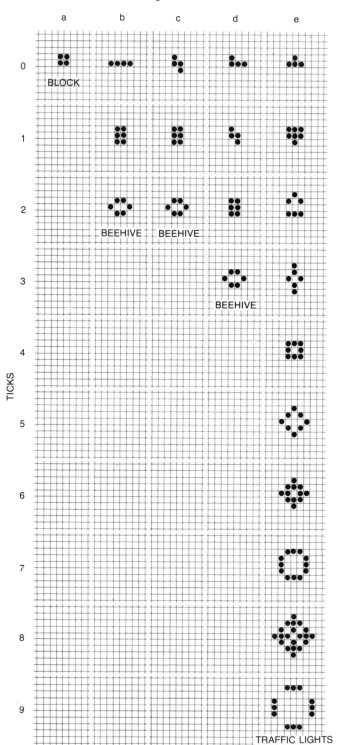

The life histories of the five tetrominoes

Figure 128

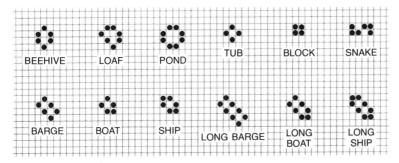

The commonest stable forms

a short time become traffic lights. The only pentomino that does not end quickly (by vanishing, becoming stable or oscillating) is the R pentomino ["a" in Figure 129]. Conway has tracked it for 460 ticks. By then it has thrown off a number of gliders. Conway remarks: "It has left a lot of miscellaneous junk stagnating around, and has only a few small active regions, so it is not at all obvious that it will continue indefinitely." Its fate is revealed in the addendum to this chapter.

Figure 129

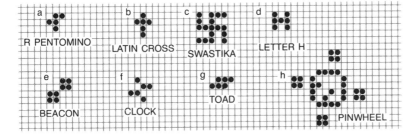

The R pentomino (a) and exercises for the reader

For such long-lived populations Conway sometimes uses a computer with a screen on which he can observe the changes. The program was written by M. J. T. Guy and S. R. Bourne. Without its help some discoveries about the game would have been difficult to make.

As easy exercises the reader is invited to discover the fate of the Latin cross ["b" in Figure 129], the swastika [c], the letter H

[d], the beacon [e], the clock [f], the toad [g] and the pinwheel [h]. The last three figures were discovered by Simon Norton. If the center counter of the H is moved up one cell to make an arch (Conway calls it "pi"), the change is unexpectedly drastic. The H quickly ends but pi has a long history. Not until after 173 ticks has it settled down to five blinkers, six blocks and two ponds. Conway also has tracked the life histories of all the hexominoes, and all but seven of the heptominoes. Some hexominoes enter the history of the R pentomino; for example, the pentomino becomes a hexomino on its first tick.

One of the most remarkable of Conway's discoveries is the five-counter glider shown in Figure 130. After two ticks it has shifted slightly and been reflected in a diagonal line. Geometers call this a "glide reflection"; hence the figure's name. After two more ticks the glider has righted itself and moved one cell diagonally down and to the right from its initial position. We mentioned earlier that the speed of a chess king is called the speed of light. Conway chose the phrase because it is the highest speed at which any kind of movement can occur on the board. No pattern can replicate itself rapidly enough to move at such speed. Conway has proved that the maximum speed diagonally is a fourth the speed of light. Since the glider replicates itself in the same orientation after four ticks, and has traveled one cell diagonally, one says that it glides across the field at a fourth the speed of light.

Figure 130

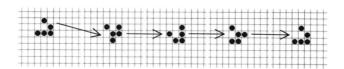

The "glider"

Movement of a finite figure horizontally or vertically into empty space, Conway has also shown, cannot exceed half the speed of light. Can any reader find a relatively simple figure that travels at such a speed? Remember, the speed is obtained by dividing the number of ticks required to replicate a figure by the number of cells it has shifted. If a figure replicates in four ticks in the same orientation after traveling two unit squares horizontally or vertically, its speed will be half that of light. Figures that move across the field by self-replication are extremely hard to find. Conway knows of four, including the glider, which he calls "spaceships" (the glider is a "feather-weight spaceship"; the others have more counters). I will disclose their patterns in the Answer Section.

Figure 131 depicts three beautiful discoveries by Conway and his collaborators. The stable honey farm [a in Figure 131] results after 14 ticks from a horizontal row of seven counters. Since a five-by-five block in one move produces the fourth generation of this life history, it becomes a honey farm after 11 ticks. The "figure 8" [b in Figure 131], an oscillator found by Norton, both resembles an 8 and has a period of 8. The form c, in Figure 131 called "pulsar CP 48–56–72," is an oscillator with a life cycle of period 3. The state shown here has 48 counters, state two has 56 and state three has 72, after which the pulsar returns to 48 again. It is generated in 32 ticks by a heptomino consisting of a horizontal row of five counters with one counter directly below each end counter of the row.

Figure 131

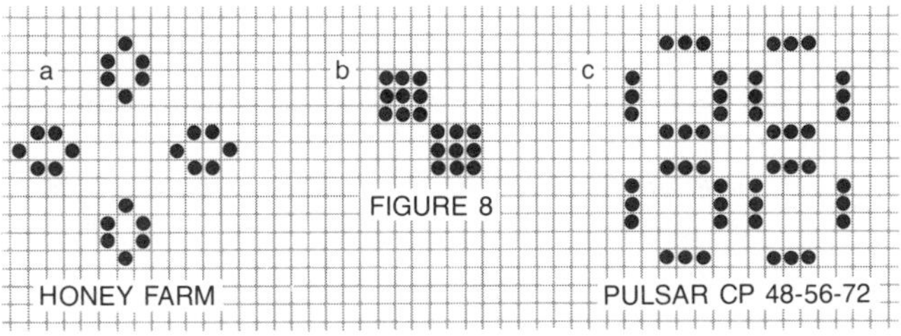

Three remarkable patterns, one stable and two oscillating

Conway has tracked the life histories of a row of n counters through $n = 20$. We have already disclosed what happens through $n = 4$. Five counters result in traffic lights, six fade away, seven produce the honey farm, eight end with four beehives and four blocks, nine produce two sets of traffic lights, and 10 lead to the "pentadecathlon," with a life cycle of period 15. Eleven counters produce two blinkers, 12 end with two beehives, 13 with two blinkers, 14 and 15 vanish, 16 give "big traffic lights" (eight blinkers), 17 end with four blocks, 18 and 19 fade away and 20 generate two blocks.

Conway also investigated rows formed by sets of n adjacent counters separated by one empty cell. When $n = 5$ the counters interact and become interesting. Infinite rows with $n = 1$ or $n = 2$ vanish in one tick, and if $n = 3$ they turn into blinkers. If $n = 4$ the row turns into a row of beehives.

The 5-5 row (two sets of five counters separated by a vacant cell) generates the pulsar *CP* 48-56-72 in 21 ticks. The 5-5-5 ends in 42 ticks with four blocks and two blinkers. The 5-5-5-5 ends in 95 ticks with four honey farms and four blinkers, 5-5-5-5-5 terminates with a spectacular display of eight gliders and eight blinkers after 66 ticks. Then the gliders crash in pairs to become eight blocks after 86 ticks. The form 5-5-5-5-5-5 ends with four blinkers after 99 ticks, and 5-5-5-5-5-5-5, Conway remarks, "is marvelous to sit watching on the computer screen." This ultimate destiny is given in the addendum.

ANSWERS

The Latin cross dies on the fifth tick. The swastika vanishes on the sixth tick. The letter *H* also dies on the sixth tick. The next three figures are flip-flops: As Conway writes, "The toad pants, the clock ticks and the beacon flashes, with period 2 in every case." The pinwheel's interior rotates 90 degrees clockwise on each move, the rest of the pattern remaining stable. Periodic figures of this kind, in which a fixed outer border is required to move the interior, Conway calls "billiard-table configurations" to distinguish them from "naturally periodic" figures such as the toad, clock and beacon.

The three known spaceships (in addition to the glider, or "featherweight spaceship" are shown in Figure 132. To be precise, each becomes a spaceship in 1 tick. (The patterns in Figure 132 never recur.) All three travel horizontally to the right with half the speed of light. As they move they throw off sparks that vanish immediately as the ships continue on their way. Unescorted spaceships cannot have bodies longer than six counters without giving birth to objects that later block their

Figure 132

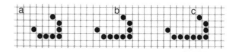

Lightweight *(left)*, middleweight *(center)*,
and heavyweight *(right)* spaceships

motion. Conway has discovered, however, that longer spaceships, which he calls "overweight" ones, can be escorted by two or more smaller ships that prevent the formation of blocking counters. Figure 133 shows a larger spaceship that can be es-

Figure 133

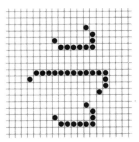

Overweight spaceship with two escorts

corted by two smaller ships. Except for this same ship, lengthened by two units, longer ships require a flotilla of more than two companions. A spaceship with a body of 100 counters, Conway finds, can be escorted safely by a flotilla of 33 smaller ships.

ADDENDUM

My 1970 column on Conway's "Life" met with such an instant enthusiastic response among computer hackers around the world that their mania for exploring "Life" forms was estimated to have cost the nation millions of dollars in illicit computer time. One computer expert, whom I shall leave nameless, installed a secret switch under his desk. If one of his bosses entered the room he would press the button and switch his computer screen from its "Life" program to one of the company's projects. The next two chapters will go into more details about the game. Here I shall comment only on some of the immediate responses to two questions left open in the first column.

The troublesome R pentomino becomes a 2-tick oscillator after 1,103 ticks. Six gliders have been produced and are traveling outward. The debris left at the center [see Figure 134] consists of four blinkers, one ship, one boat, one loaf, four beehives, and eight blocks. This was first established at Case Western Reserve University by Gary Filipski and Brad Morgan, and later confirmed by scores of "Life" hackers here and abroad.

The fate of the 5-5-5-5-5-5-5 was first independently found by Robert T. Wainwright and a group of hackers at Honeywell's Computer Control Division, later by many others. The pattern stabilizes as a 2-tick oscillator after 323 ticks with four traffic lights, eight blinkers, eight loaves, eight beehives, and

Figure 134

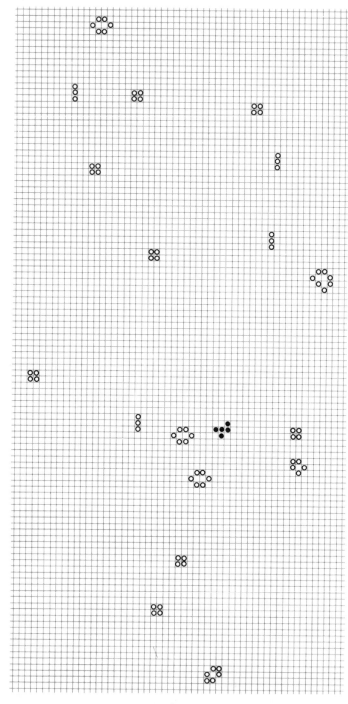

R pentomino's original (black) and final (open dots) state.
(Six gliders are out of sight.)

four blocks. Figure 135 reproduces a printout of the final steady state. Because symmetry cannot be lost in the history of any life form, the vertical and horizontal axes of the original symmetry are preserved in the final state. The maximum population (492 bits) is reached in generation 283, and the final population is 192.

Figure 135

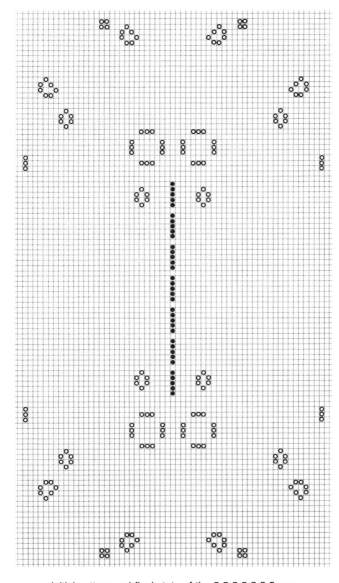

Initial pattern and final state of the 5-5-5-5-5-5-5 row

21

THE GAME OF LIFE, PART II

Cellular automata theory began in the mid-fifties when John von Neumann set himself the task of proving that self-replicating machines were possible. Such a machine, given proper instructions, would build an exact duplicate of itself. Each of the two machines would then build another, the four would become eight, and so on. (This proliferation of self-replicating automata is the basis of Lord Dunsany's amusing 1951 novel *The Last Revolution*.) Von Neumann first proved his case with "kinematic" models of a machine that could roam through a warehouse of parts, select needed components and put together a copy of itself. Later, adopting an inspired suggestion by his friend Stanislaw M. Ulam, he showed the possibility of such machines in a more elegant and abstract way.

Von Neumann's new proof used what is now called a "uniform cellular space" equivalent to an infinite checkerboard. Each cell can have any finite number of "states," including a "quiescent" (or empty) state, and a finite set of "neighbor" cells that can influence its state. The pattern of states changes in discrete time steps according to a set of "transition rules" that apply simultaneously to every cell. The cells symbolize the basic parts of a finite-state automaton and a configuration of live cells is an idealized model of such a machine. Conway's game of "Life" is based on just such a space. His neighborhood consists of the eight cells surrounding a cell; each cell has two states (empty or filled), and his transition rules are the birth, death and survival rules I explained in the previous chapter. Von Neumann, applying transition rules to a space in which each cell has 29 states and four orthogonally adjacent neighbors, proved the existence of a configuration of about 200,000 cells that would self-reproduce.

The reason for such an enormous configuration is that, for von Neumann's proof to apply to actual automata, it was necessary that his cellular space be capable of simulating a Turing machine: an idealized automaton, named for its inventor, the British mathematician A. M. Turing, capable of performing any desired calculation. By embedding this universal computer in his configuration, von Neumann was able to produce a universal constructor. Because it could in principle construct any desired configuration by stretching "arms" into an empty region of the cellular space, it would self-replicate when given a blueprint of itself. Since von Neumann's death in 1957 his existence proof (the actual configuration is too vast to construct and manipulate) has been greatly simplified. The latest and best reduction, by Edwin Roger Banks, a mechanical engineering graduate student at the Massachusetts Institute of Technology, does the job with cells of only four states.

Self-replication in a trivial sense—without using configurations that contain Turing machines—is easy to achieve. A delightfully simple example, discovered by Edward Fredkin of M.I.T. about 1960, uses two-state cells, the von Neumann neighborhood of four orthogonally adjacent cells and the following parity rule: Each cell with an even number of live neighbors (0, 2, 4) at time t becomes or remains empty at time $t+1$, and each cell with an odd number of neighbors (1, 3) at time t becomes or remains live at time $t+1$. It is not hard to show that after 2^n ticks (n varying with different patterns) any initial pattern of live cells will reproduce itself four times— above, below, left and right of an empty space that it formerly occupied. The four replicas will be displaced 2^n cells from the vanished original. The new pattern will, of course, replicate again after another 2^n steps, so that the duplicates keep quadrupling in the endless series 1, 4, 16, 64, ... Figure 136 shows two quadruplings of a right tromino. Terry Winograd, in a 1967 term paper written when he was an M.I.T. student, generalized Fredkin's rule to other neighborhoods, any number of dimensions and cells with any prime number of states.

Ulam investigated a variety of cellular automata games, experimenting with different neighborhoods, numbers of states and transition rules. In a 1967 paper "On Recursively Defined Geometrical Objects and Patterns of Growth," written with Robert G. Schrandt, Ulam described a number of different games. Figure 137 shows generation 45 of a history that began with one counter on the central cell. As in Conway's game, the cells are two-state, but the neighborhood is that of von Neumann (four adjacent orthogonal cells). Births occur on cells

Figure 136

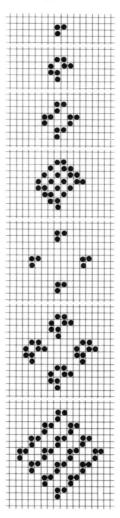

The replication of a tromino

that have one and only one neighbor, and all live cells of generation n vanish when generation $n+2$ is born. In other words, only the last two generations survive at any step. In Figure 137 the 444 new births are shown as black cells. The 404 white cells of the preceding generation will all disappear on the next tick. Note the characteristic subpattern, which Ulam calls a "dog bone." Ulam experimented with games in which two configurations were allowed to grow until they collided. In the ensuing

Figure 137

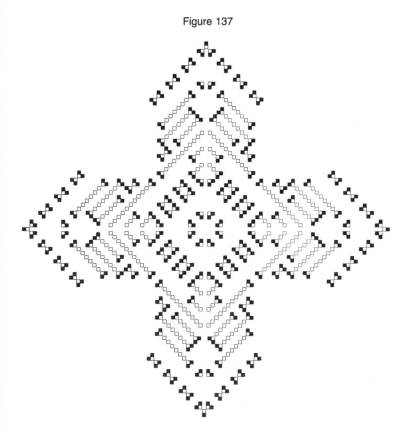

Generation 45 in a cellular game devised by
Stanislaw M. Ulam

"battle" one side would sometimes wipe out the other; sometimes both armies would be annihilated. Ulam also explored
games on three-dimensional cubical tessellations. His major papers on cellular automata are in *Essays on Cellular Automata*, edited by Arthur W. Burks.

Similar games can be devised for triangular and hexagonal
tessellations but, although they *look* different, they are not essentially so. All can be translated into equivalent games on a
square tessellation by a suitable definition of "neighborhood."
A neighborhood need not be made up of touching cells. In
chess, for instance, a knight's neighborhood consists of the
squares to which it can leap and squares on which there are
pieces that can attack it. As Burks has pointed out, games such
as chess, checkers and go can be regarded as cellular automata
games in which there are complicated neighborhoods and tran-

sition rules and in which players choose among alternative next states in an attempt to be first to reach a certain final state that wins.

Among the notable contributions of Edward F. Moore to cellular automata theory the best-known is a technique for proving the existence of what John W. Tukey named "Garden of Eden" patterns. These are configurations that cannot arise in a game because no preceding generation can form them. They appear only if given in the initial (zero) generation. Because such a configuration has no predecessor, it cannot be self-reproducing. I shall not describe Moore's ingenious technique because he explained it informally in an article in *Scientific American* (see "Mathematics in the Biological Sciences," by Edward F. Moore; September, 1964) and more formally in a paper that is included in Burks's anthology.

Alvy Ray Smith III, a cellular automata expert at New York University's School of Engineering and Science, found a simple application of Moore's technique to Conway's game. Consider two five-by-five squares, one with all cells empty, the other with one counter in the center. Because, in one tick, the central nine cells of both squares are certain to become identical (in this case all cells empty) they are said to be "mutually erasable." It follows from Moore's theorem that a Garden of Eden configuration must exist in Conway's game. Unfortunately the proof does not tell how to find such a pattern and so far none is known. It may be simple or it may be enormously complex. Using one of Moore's formulas, Smith has been able to calculate that such a pattern exists within a square of 10 billion cells on a side, which does not help much in finding one.

Smith has been working on cellular automata that simulate pattern-recognition machines. Although this is now only of theoretical interest, the time may come when robots will need "retinas" for recognizing patterns. The speeds of scanning devices are slow compared with the speeds obtainable by the "parallel computation" of animal retinas, which simultaneously transmit thousands of messages to the brain. Parallel computation is the only way new computers can increase significantly in speed because without it they are limited by the speed of light through miniaturized circuitry. The cover of the February, 1971, issue of *Scientific American* [reproduced in Figure 138] shows a simple procedure, devised by Smith, by which a finite one-dimensional cellular space employs parallel computation for recognizing palindromic symmetry. Each cell has many possible states, the number depending on the number of different symbols in the palindrome, and a cell's neighborhood is the two cells on each side.

Figure 138

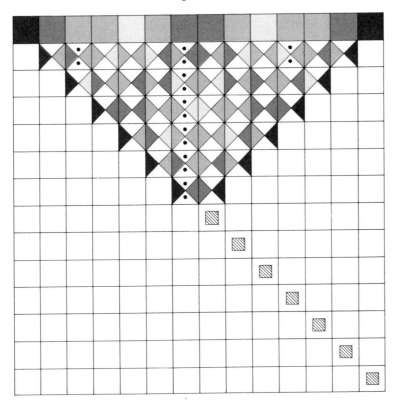

Cellular automaton

Smith symbolizes the palindrome TOO HOT TO HOOT with four states of cells in the top row. *T, O* and *H* are represented by blue, red and yellow respectively, and black marks the palindrome's two ends. Here we have indicated the colors by different shadings. The white cells in the other rows are in the quiescent state. The horizontal rows below the top row are successive generations of the top configuration when certain transition rules are followed in discrete time steps. In other words, the picture is a space-time diagram of a single row, each successive row indicating the next generation.

In the first transition each shade travels one cell to the left and one cell to the right, except for the end shadings, which are blocked by black; black moves inward at each step. Each cell on which two shadings land acquires a new state, symbolized by a cell divided into four triangles. The left triangle has the shading that was previously on the left, the right triangle has the shading previously on the right. The result of this first move is shown in the second row. When an adjacent pair of cells forms a tilted square in the center that is a solid shading, it indicates a "collision" of like shadings and is symbolized by black dots in the two white triangles of the left cell. Dots remain in that cell for all subsequent generations unless a collision of unlike shadings occurs to the immediate right of the dotted cell, in which case the dots are erased. When collisions of *unlike* shadings occur, the left cell of the pair remains undotted for all subsequent generations even though like shadings may later collide on its right.

At each move the shadings continue to travel one cell left or right (the direction in which the shaded triangles point) and all rules apply. If the palindrome has n letters, with n even as in this example (the scheme is modified slightly if n is odd), it is easy to see that after $n/2$ moves only two adjacent nonquiescent cells remain. If the left cell of this pair is dotted, the automaton has recognized the initial row as being palindromic. Down the diagram's center you see the colliding pairs of like shadings in the same order as they appear on the palindrome from the center to each end. As soon as recognition occurs the left cell of the last pair is erased and the right cell is altered to an "accept" state, here symbolized by nested squares. An undotted left cell would signal a nonpalindrome, in which case the left cell would become blank and the right cell would go into a "reject" state.

A Turing machine, which computes serially, requires in general n^2 steps to recognize a palindrome of length n. Although recognition occurs here at step $n/2$, the accept state is shown moving in subsequent generations to the right to symbolize the cell-by-cell transmission of the acceptance to an output boundary of the cellular space. Of course it is easy to construct more efficient palindrome-recognizing devices with actual electronic hardware, but the point here is to do it with a highly abstract, one-dimensional cellular space in which information can pass only from a cell to adjacent cells and not even the center of the initial series of symbols is known at the outset. As Smith puts it anthropomorphically, after the first step each of the three dotted cells thinks it is at the center of a palindrome. The dotted cells at each end are disillusioned on the next move because of the collision of unlike shading₋ at their right. Not until generation $n/2$ does the dotted cell at the center know it *is* at the center.

Now for some startling new results concerning Conway's game. Conway was fully aware of earlier games and it was with them in mind that he selected his recursive rules with great care to avoid two extremes: too many patterns that grow

quickly without limit and too many that fade quickly. By striking a delicate balance he designed a game of surprising unpre-
dictability and one that produced such remarkable figures as
oscillators and moving spaceships. He conjectured that no finite population could grow (in number of members) without
limit, and he offered $50 for the first proof or disproof. The
prize was won in November, 1970, by a group in the Artificial
Intelligence Project at M.I.T. consisting of (in alphabetical order) Robert April, Michael Beeler, R. William Gosper, Jr.,
Richard Howell, Rich Schroeppel and Michael Speciner. Using
a program devised by Speciner for displaying life histories on
an oscilloscope, Gosper made a truly astounding discovery: he
found a glider gun! The configuration in Figure 139 grows
into such a gun, firing its first glider on tick 40. The gun is an
oscillator of period 30 that ejects a new glider every 30 ticks.
Since each glider adds five more counters to the field, the population obviously grows without limit.

Figure 139

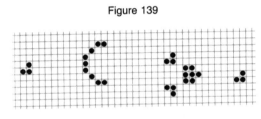

A configuration that grows into a glider gun

The glider gun led the M.I.T. group to many other amazing
discoveries. A series of printouts (supplied by Robert T. Wainwright of Yorktown Heights, N.Y.) shows how 13 gliders crash
to form a glider gun [see Figure 140]. The last five printouts
show the gun in full action. The group also found a way to position a pentadecathlon [see Figure 141], an oscillator of period
15, so that it "eats" every glider that strikes it. A pentadecathlon can also reflect a glider 180 degrees, making it possible for
two pentadecathlons to shuttle a glider back and forth forever.
Streams of intersecting gliders produce fantastic results. Strange
patterns can be created that in turn emit gliders. Sometimes
collision configurations grow until they ingest all guns. In other
cases the collision mass destroys one or more guns by shooting
back. The group's latest burst of virtuosity is a way of placing
eight guns so that the intersecting streams of gliders build a
factory that assembles and fires a middleweight spaceship
about every 300 ticks.

Figure 140

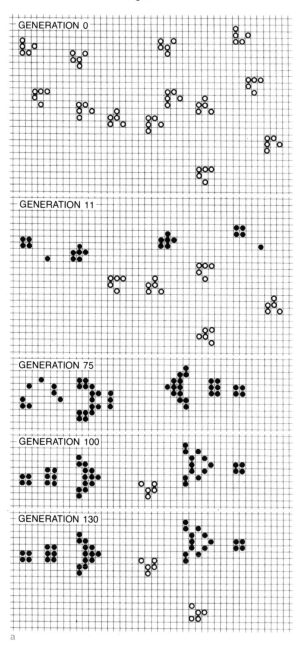

Here and on the facing page 13 gliders crash to
form a glider gun (generation 75) that oscillates
with a period of 30, firing a glider in each cycle

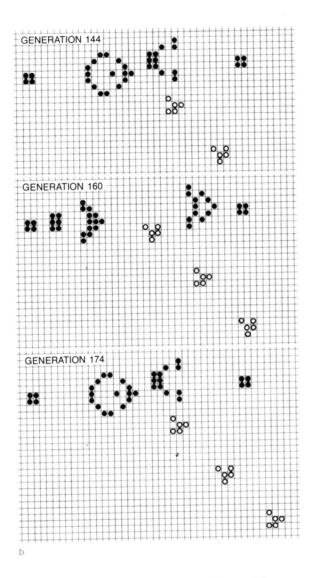

b

The existence of glider guns raises the exciting possibility that Conway's game will allow the simulation of a Turing machine, a universal calculator capable in principle of doing anything the most powerful computer can do. The trick would be to use gliders as unit pulses for storing and transmitting information and performing the required logic operations that are handled in actual computers by their circuitry. If Conway's game allows a universal calculator, the next question will be

Figure 141

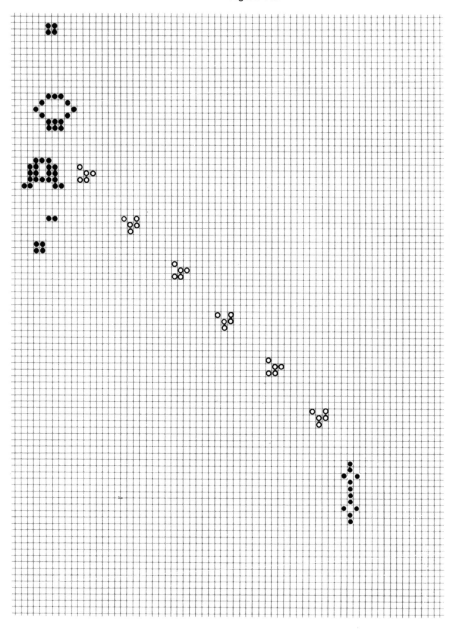

Pentadecathlon *(bottom right)* "eats" gliders
fired by the gun

whether it allows a universal constructor, from which nontrivial self-replication would follow. So far this has not been achieved with a two-state space and Conway's neighborhood, although it has been proved impossible with two states and the von Neumann neighborhood.

The M.I.T. group found many new oscillators [see *Figure* 142]. One of them, the barber pole, can be stretched to any length and is a flip-flop, with each state a mirror image of the other. Another, which they rediscovered, is a pattern Conway's group had found earlier and called a Hertz oscillator. Every four ticks the hollow "bit" switches from one side of the central frame to the other, making it an oscillator of period 8. The tumbler, which was found by George D. Collins, Jr., of McLean, Va., turns upside down every seven ticks.

Figure 142

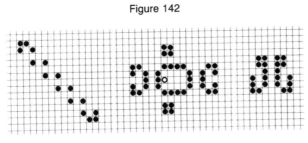

Barber pole *(left)*, Hertz oscillator *(middle)*, and tumbler *(right)*

The Cheshire cat [see *Figure* 143] was discovered by C. R. Tompkins of Corona, Calif. On the sixth tick the face vanishes, leaving only a grin; the grin fades on the next tick and only a permanent paw print (block) remains. The harvester was constructed by David W. Poyner of Basildon in England. It plows up an infinite diagonal at the speed of light, oscillating with period 4 and ejecting stable packages along the way [see *Figure* 144]. "Unfortunately," writes Poyner, "I have been unable to develop a propagator that will sow as fast as the harvester will reap."

Wainwright has made a number of intriguing investigations. He filled a 120-by-120 square field with 4,800 randomly placed bits (a density of one-third) and tracked their history for 450 generations, by which time the density of this primordial soup, as Wainwright calls it, had thinned steadily to one-sixth.

Figure 143

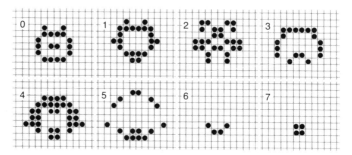

The Cheshire cat (0) fades to a grin (6)
and disappears, leaving a paw print (7)

Figure 144

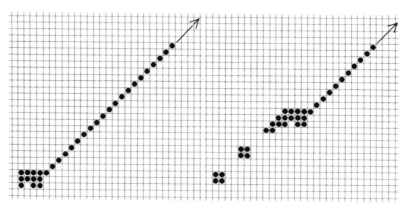

The harvester, shown at generations (0) *left*
and 10 *(right)*

Whether it would eventually vanish or, as Wainwright says, percolate at a constant minimum density is anybody's guess. At any rate, during the 450 generations 42 short-lived gliders were formed. Wainwright found 14 different patterns that became glider states on the next tick. The pattern that produced the greatest number of gliders (14 in all) is shown [*a in Figure* 145]. A Z-pattern found by Collins and by Jeffrey Lund of Pewaukee, Wis., after 12 ticks becomes two gliders that sail off in opposite directions [*b in Figure* 145]. Wainwright and others set two gliders on a collision course that causes all bits to vanish on the fourth tick [*c in Figure* 145]. Wallace W. Wagner of Ana-

Figure 145

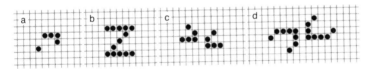

Two spawners of gliders and two collision courses

heim, Calif., found a collision course for two lightweight space-ships that also ends (on the seventh tick) in total blankness [*d in Figure* 145].

Wainwright has experimented with various infinite fields of regular stable patterns, which he calls agars—rich culture me-diums. When, for instance, a single "virus," or bit, is placed in the agar of blocks shown in Figure 146 so that it touches the corners of four blocks, the agar eliminates the virus and re-pairs itself in two ticks. If, however, the alien bit is positioned as shown (or at any of the seven other symmetrically equivalent spots), it initiates an inexorable disintegration of the pattern. The portion eaten away contains active debris that has overall bilateral symmetry along one axis and a roughly oval border that expands, probably forever, in the four compass directions at the speed of light.

Figure 146

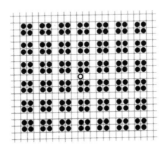

Agar doomed by a virus

The most immediate practical application of cellular auto-mata theory, Banks believes, is likely to be the design of circuits capable of self-repair or the wiring of any specified type of new circuit. No one can say how significant the theory may eventu-ally become for the physical and biological sciences. It may have important bearings on cell growth in embryos, the replication of DNA molecules, the operation of nerve nets, genetic changes in evolving populations and so on. Analogies with life processes are impossible to resist. If a primordial broth of amino acids is large enough, and there is sufficient time, self-replicating, moving automata may result from complex transition rules built into the structure of matter and the laws of nature. There is even the possibility that space-time itself is granular, composed of discrete units, and that the universe, as Fredkin and others have suggested, is a vast cellular automaton run by an enormous computer. If so, what we call motion may be only simulated motion. A moving spaceship, on the ultimate microlevel, may be essentially the same as one of Conway's spaceships, appearing to move on the macrolevel whereas actually there is only an alteration of states of basic space-time cells in obedience to transition rules that have not yet been discovered.

22

THE GAME OF LIFE, PART III

So much has been discovered about Conway's "Life" since I first wrote the last two chapters, that it was impossible to summarize the highlights in an addendum. A book could and should be written about the game, an *Encyclopedia of Life*, or a *Handbook of Life*, that would put all the important known Life forms on record and thereby save Lifenthusiasts the labor of rediscovering them. The eleven issues that appeared of Robert Wainwright's periodical *Lifeline* continue to be the main repository of such data. Wainwright is said to be working on a book, and there are rumors of other books about "Life" that are in the making. In the meantime, I will try in this chapter to pull together some of the significant developments in "Life" since my second column on the game ran in *Scientific American* in 1971. Because so many basic forms were independently discovered by many people, I shall not often attempt to credit first discoverers.

The earliest and most important group of Lifenthusiasts was at M.I.T., centering around William Gosper who is now working for Xerox at their Stanford research headquarters. In the mid-70s the most active "Life" group was in the computer control division of Honeywell, Inc., Framington, Mass. It included (alphabetical order) Thomas Holmes, Keith McClelland, Michael Sporer, Philip Stanley, Donald Woods, and his father William Woods. In the late seventies, an active group of "Life" hackers formed at the University of Waterloo, in Canada, with John Abbott, David Buckingham, Mark Niemiec, and Peter Raynham as the leaders. Most of what I shall report comes from these three groups.

All still lifes with 13 or fewer bits have long been known. The block and tub are the only 4-bit stable forms, and the boat

Figure 147

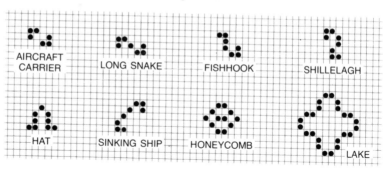

More still lifes

is the only one with 5 bits. Figure 128 caught four of the five 6-bit still lifes, missing only the aircraft carrier shown in Figure 147. There are four 7-bit stable forms: the loaf, long boat, long snake, and fishhook. The fishhook or "eater" is the smallest still life lacking any kind of symmetry. Note that forms such as the boat, barge, ship, and sinking ship can be stretched to any length, and lakes can be made as large as you like, with any number of barges, boats, and ships at anchor on the water. There are nine 8-bit still lifes, ten 9-bit forms, 25 with 10 bits, 46 with 11 bits, 121 with 12 bits, and 149 with 13 bits. The stable pool table in Figure 148 was constructed out of long sinking ships and parts of ponds by William Woods.

Figure 148

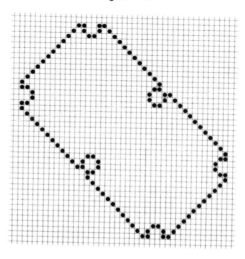

The stable pool table

Figure 149

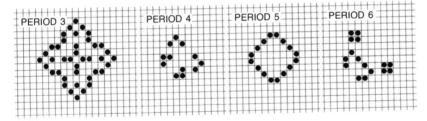

Low-period oscillators

Hundreds of elegant oscillators have been found. Figure 149 shows a few of small size, with short periods. The M.I.T. group, early in the history of "Life," found easy ways to construct giant flip-flops (period-2 oscillators) such as the one shown in Figure 150. It oscillates between the patterns shown in black dots and circles.

Figure 150

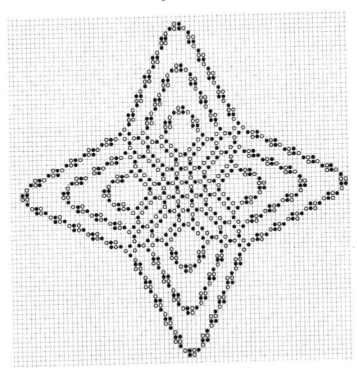

A flip-flop pattern that alternates between states
shown in black and with circles

Another large class of "Life" forms that have been intensively investigated are what the Honeywell group named the fuses. These are stems one or more bits wide, either diagonal or orthogonal, usually infinite in length, that burn steadily from one end toward the other. The simplest is the fuse shown in Figure 151 *a,* a diagonal of bits that either rises to infinity or has a stable top as shown. It simply burns itself out without producing any sparks or stable smoke. If you put another bit to the left of the lower end, it forms a tiny flame that travels along with the burning.

Figure 151

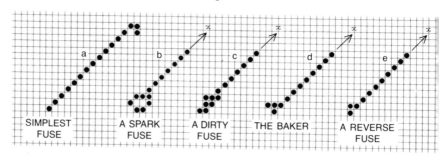

Five fuses

Fuse *b* in Figure 151 oscillates with a period of 4, giving off sparks that fade quickly. A "dirty fuse," like the one shown in *c* in Figure 151, leaves clouds of debris behind as it burns. At one point it shoots off a glider. Fuse *d* in Figure 151, named the "baker" by its discoverer, McClelland, is a confused fuse that bakes a string of stable loaves while it burns. The last three fuses all oscillate with periods of 4, and all four burn with the speed of light.

Fuse *e* in Figure 151, eventually becomes a clean fuse of period 4, but leaves behind a cloud consisting of three blocks, three beehives, two blinkers, a ship, and four gliders. William Woods calls it a "reverse fuse" because it explodes first, then burns quietly for the rest of its endless life. The harvester, described in the previous chapter, is of course a fuse.

Other unusual fuses are shown in Figure 152. Fuse *a,* found by Steve Tower, has a period of 8. It leaves behind a trail of beacons. Fuse *b* abandons a twin pair of boats every four ticks. Orthogonal fuse *c,* which burns with a speed slower than light, consumes two tubs every 18 ticks, then changes them to traffic lights (four blinkers). It was discovered by Earl Abbe. Wain-

Figure 152

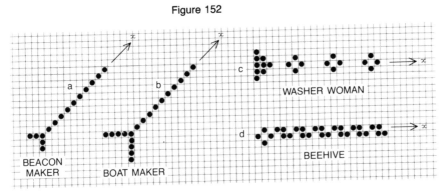

More fuses

wright's fuse *d* consumes three fenceposts every 12 generations, and turns them into a beehive.

Two fuses of a more complicated nature, discovered by Don
Woods, are shown in Figure 153. The cow burns at light speed,
with period 8, slowly "chewing its cud" by eating the blocks on
either side, bringing them back again, then eating them a second time. The two-glider fuse throws off two gliders every 12
ticks. I resist the impulse to describe two close relatives of
fuses, the wicks (infinite in both directions) and the kinkbombs.
Kinkbombs come in three varieties: duds, firecrackers, and
bombs, as detailed by Mark Horton in the 11th issue of *Lifeline*.

Figure 153

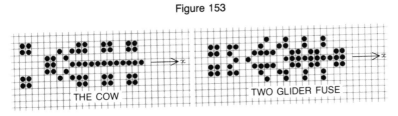

Two remarkable fuses

There are 102 distinct patterns of bits within a 3 × 3 square
(excluding rotations and reflections, but including the patterns
consisting of nine bits and no bits). Some of these are polyominoes, some not. All the letters of the alphabet in Braille are
among the 102. The fates of all 102 are known. Also known
are the fates of all polyominoes through the order-7
heptominoes.

Methuselah patterns are those of fewer than 10 bits which do not stabilize until after more than 50 generations. Two examples were given in the previous chapter: The 5-bit R-pentomino and the pi-heptomino of 7 bits. The first generation of the pi-heptomino, by the way, reappears in tick 31, but shifted 9 cells. Because of interaction with its exhaust, in generation 61, it fails to make it as a spaceship.

Other examples of Methuselahs are shown in Figure 154. The first one, a is the smallest known. It becomes the R-pentomino in two ticks, giving it a life of 1,105 generations. Methuselah b stabilizes (six blocks, twelve blinkers, one loaf) after 608 generations, c (the thunderbird) lasts 243 ticks, and d goes to 1,108. The heptomino e stabilizes after 148 ticks, having produced three blocks, a ship, and two gliders. The acorn f, found by Charles Corderman, is the most amazing Methuselah known. It lives for 5,206 generations! When it stabilizes as an "oak" of 633 bits, it has produced numerous gliders, 13 of which escape.

Figure 154

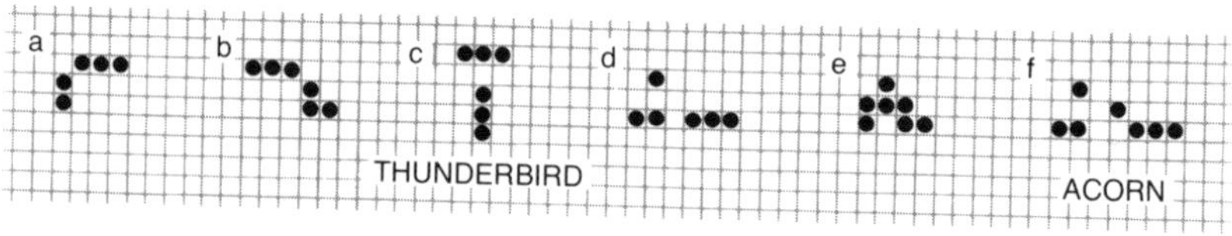

Methusalehs

The Honeywell group tracked the life histories of the first nine members of the 5-cell crosses, of which the simplest are shown in Figure 155. The first is a portion of an infinite trellis

Figure 155

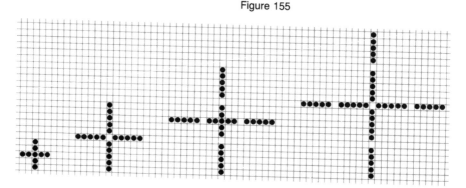

The five-cell cross series

consisting of solid horizontal and vertical rows, two cells apart, that surround an infinity of empty 2×2 squares. Like the infinite trellis, this cross vanishes in one tick. The next cross disappears in 8 ticks. The third ends with many traffic lights in 6 ticks, and the fourth stabilizes after 34 ticks with eight blinkers, having produced a truly spectacular display of fireworks along the way. (Its 19th generation is a beautiful ring of blocks with a checkerboard in the center.) Order-5 and order-7 crosses in this sequence stabilize as four pulsars in 36 and 21 ticks respectively, orders 6 and 8 go to four pulsars and a tub in 36 and 21 ticks respectively, and order-9 ends after 42 ticks with 16 blocks and 8 blinkers.

William Gosper, in 1971, found the eater (fishhook), the incredible 7-bit stable form shown with circles in Figure 156. It has the ability to consume an enormous variety of "Life" forms, then quickly repair itself. The first four pictures show the eater about to ingest a glider, blinker, pre-beehive, and a lightweight spaceship. In the fifth picture two eaters are poised to devour one another. This is prevented by their amazing ability to self-repair, so the pattern oscillates with period 3. The last picture shows how two gliders collide to produce an eater on the 13th tick. In recent years eaters of larger size have been discovered, with a variety of bizarre feeding habits.

Figure 156

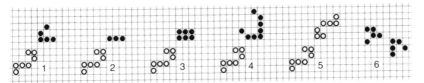

The eater (circles) and some of its prey

Extensive investigations have been made of different kinds of agars (regular patterns that are infinite in two dimensions), the procrastinators (forms that take more than 50 ticks to become a single simple stable form), and puffer trains. The puffers leave a trail of permanent smoke. Three are shown in Figure 157. The first, discovered by Gosper, is an engine escorted between two lightweight spaceships. It puffs along at half the speed of light until after more than 1,000 ticks it develops a period of 140. Paul Schick discovered an entire family of puffer trains, the simplest of which, shown in *b*, leaves nothing behind. The pair of mirror-image lightweight spaceships pull along the symmetrical heptomino engine with a period of

Figure 157

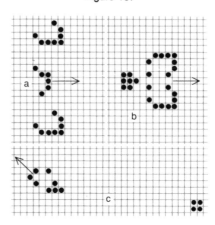

Puffer trains

12. The switch-engine puffer train *c* in Figure 157, moves too slowly (one-twelfth the speed of light) to be of much use. It travels diagonally like a glider, eventually producing eight blocks every 288 generations. No escorting spaceships are needed, but without the stabilizing block its smoke catches up with the engine and destroys it.

The first Garden of Eden pattern, reproduced in Figure 158, was found by Roger Banks in 1971. It required an enormous computer search of all possible predecessor patterns. The confining rectangle (9×33) holds 226 bits. The only other Garden of Eden pattern known was found by a French group in 1974, led by J. Hardouin-Duparc, at the University of Bordeaux. It is inside a rectangle of 6×122.

Figure 158

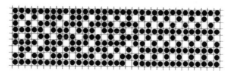

A garden of Eden

Although any "Life" pattern generates only one successor, the converse is not true. A given pattern may have two or more predecessors. This is why searching for Garden of Eden patterns is so difficult—the computer has to look at all possible

predecessors at each backward tick. If the universe eventually
turns out to be one monstrous cellular automaton, one may
reasonably ask whether there is an initial Garden of Eden state
that required a creation because it has no predecessor pattern.
By the way, the fact that a "son" of a Garden of Eden pattern
may have more than one "father" has led Conway to offer $50
to the first person who can find a pattern that has a father but
no grandfather. The existence of such a pattern is still an open
question.

The most spectacular of the new developments in "Life" involve gliders and their collisions. Gosper's group found new
types of glider guns, more compact spaceship factories produced by glider crashes, and innumerable "Life" forms that eat
gliders or reflect them back at different angles. Before its
members broke up to go their separate ways, the M.I.T. group
managed to complete a 17-minute film about their discoveries
that has become a classic.

A pure glider generator is one that generates one or more
gliders with no debris left over. Two elegant ones found by the
Honeywell group are shown in Figure 159. The biloaf *left* in
four ticks produces two gliders going opposite ways. The 4-8-
12 diamond *right* in 15 ticks forms four gliders headed in four
different directions. Half a dozen 5-bit forms turn into a single
glider, and more than a hundred 6-bit forms do the same. A
search for predecessors of the original Gosper glider gun
turned up a pattern of 21 bits that is the smallest known,
though it seems possible there may be a way of positioning just
four gliders (20 bits) so that they crash and form a gun.

Figure 159

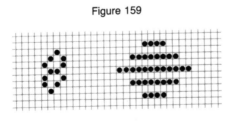

Two glider-generators

I mentioned earlier Gosper's way of placing eight guns so
that their gliders crash to form a spaceship factory which fires
off a middleweight spaceship about every 300 generations.
Gosper soon improved this to four guns and one pentadecathlon. This pattern produces a factory that shoots off lightweight
or middleweight spaceships (depending on the timing) every

60 ticks. Wainwright positioned three "newguns" that generate a middleweight spaceship every 46 generations.

Lifenthusiasts have investigated thousands of ways that gliders and spaceships can collide to produce an incredible variety of stable patterns (including the null pattern of nothing at all), as well as patterns that change, and patterns that produce new gliders and/or spaceships. Figure 160 shows some unusual collisions found by the Waterloo group. On the left is the pattern just before the crash; on the right, the outcome after the indicated number of ticks (t = ticks).

The breeder is one of the most remarkable life forms found by the M.I.T. group; remarkable because its population growth is so rapid. Figure 161 is a photograph of a computer scope that shows the breeder breeding gliders. The little dots are gliders, about 1,000 of them inside the triangular region. The breeder consists of ten puffer trains moving east, their exhaust carefully controlled so that they generate gliders that crash to form guns that instantly spring into action along the horizontal axis. The picture shows the breeder at generation 3,333. Thirty guns are firing northeast at a rate of one glider per tick. The firing rate increases without limit until at about tick 6,500 the number of gliders starts to exceed the age of the breeder. Seeing the breeder in action was one of the most awesome high points of my visit to M.I.T.

When I wrote the previous chapter for the February 1971 issue of *Scientific American*, I raised the question of whether the rules of "Life" permit the construction of a universal computer. I had the pleasure of reporting the next month that "Life" is indeed universal. Gosper at M.I.T. and Conway at Cambridge independently "universalized" the "Life" space by showing how gliders could be used as pulses to simulate a Turing machine. Exactly how this is done is too complicated to go into here, but you will find it clearly outlined by Conway in the second volume of *Winning Ways*, the book he coauthored with Elwyn Berlekamp and Richard Guy.

The universality of "Life" means that it is possible in principle to use moving gliders to perform any calculation that can be performed by the most powerful digital computer. For example, one can arrange a formation of glider guns, eaters, and other "Life" forms so that a stream of gliders, with gaps in the right places, will calculate pi, e, the square root of 2, or any other real number to any number of decimal places. Of course, it would be a very inefficient way to do such calculations, nonetheless they are possible if you have a large enough field and sufficient ingenuity to build the needed "machine."

In *Winning Ways* Conway uses Fermat's last theorem to illus-

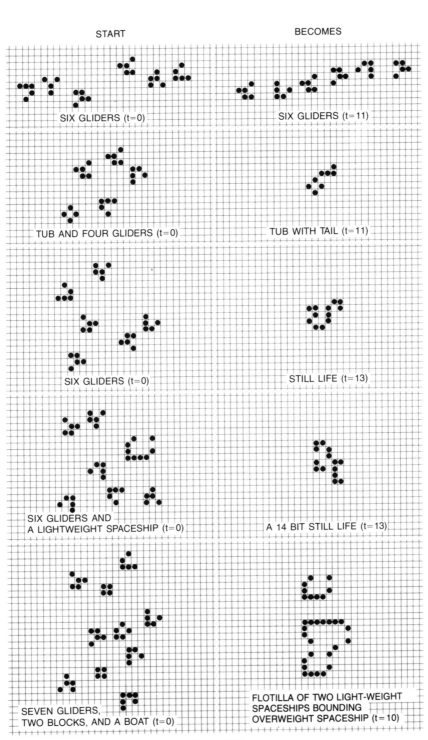

START   BECOMES

SIX GLIDERS (t=0)   SIX GLIDERS (t=11)

TUB AND FOUR GLIDERS (t=0)   TUB WITH TAIL (t=11)

SIX GLIDERS (t=0)   STILL LIFE (t=13)

SIX GLIDERS AND
A LIGHTWEIGHT SPACESHIP (t=0)   A 14 BIT STILL LIFE (t=13)

SEVEN GLIDERS,
TWO BLOCKS, AND A BOAT (t=0)

FLOTILLA OF TWO LIGHT-WEIGHT
SPACESHIPS BOUNDING
OVERWEIGHT SPACESHIP (t=10)

Figure 160

Figure 161

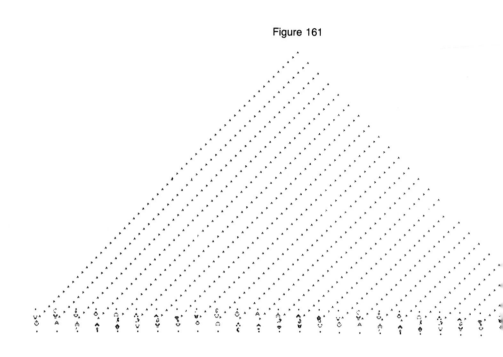

The breeder

trate "Life's" computing power as well as its limitations. A "Life" machine can be constructed that will steadily test the values of the four variables in Fermat's famous formula. The program could be designed to halt, say by fading away, if it found a counterexample to Fermat's conjecture. On the other hand, if the conjecture is true, the "Life" machine will keep searching forever for the right combination of values. We know from undecidability theory that there is no way to know in advance whether any given problem is solvable, therefore there is no way to know in advance whether any given pattern in "Life" will continue to change or to reach a stable end.

In 1981, in a letter telling me he had found "Life" to be universal, Conway added a note on the back of the envelope. "If

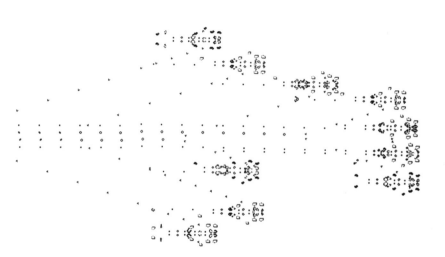

(ask Gosper) gliders can crash to form a pentadecathlon, then I can produce self-replicating machines, and it's undecidable whether a given machine is self-replicating."

I cannot remember if I asked Gosper this question, but at any rate, gliders *can* crash to form pentadecathlons, and Conway states flatly, in *Winning Ways,* that self-replicating machines can be constructed in "Life" space. We are not speaking now of moving forms like spaceships, but of machines that will build exact copies of themselves. The original machine may either remain in the space or it can be programmed to self-destruct after it has replicated itself. So far as I know no one has built such a machine, but if Conway is right (his proof has not been published), it is possible to build them.

Conway also asserts in *Winning Ways* that he has proved that "Life" patterns exist which move steadily in any desired rational direction, recovering their initial forms after a fixed number of moves. As for spaceships (which move without producing smoke), no new ones have been discovered other than those already known to Conway in 1970.

Conway goes on to speculate that if you imagine a sufficiently large broth of randomly placed bits, one would expect that by pure chance self-replicating creatures would arise, and those best adapted to survive would live longer than the others. Interactions with the environment would produce mutations. As in organic evolution, most mutations would be harmful, but some would have survival value. "It's probable," Conway writes, "given a large enough "Life" space, initially in a random state, that after a long time, intelligent self-reproducing animals will emerge and populate some parts of the space."

I would prefer the word "possible" here to "probable," but there is no question that "Life's" analogy with biological evolution on earth is remarkable. One science fantasy writer, the widely read Piers Anthony, plays with this theme in his 1976 novel, *Ox*. Diagrams of "Life" patterns head each chapter, and the book's plot involves intelligent, sentient beings called "pattern entities" or "sparkle clouds" that have evolved by just the process Conway imagines, in a cellular space of dimensions higher than our spacetime. Their behavior is totally determined by transition rules, but like us they imagine themselves to have free wills. There is an amusing Chapter 11 in which Cal explains the rules of "Life" to Aquilon and she experiments with some simple patterns.

"Try this one," Cal suggests, giving her the *R*-pentomino:

> "That's similar to the one I just did. You've just tilted it sideways, which makes no topological difference, and added one dot."
>
> "Try it," he repeated.
>
> She tried it, humoring him. But soon it was obvious that the solution was not a simple one. Her numbered patterns grew and changed, taking up more and more of the working area. The problem ceased to be merely intriguing; it became compulsive. Cal well understood this; he had been through it himself. She was oblivious to him now, her hair falling across her face in attractive disarray, teeth biting lips. "What a difference a dot makes!" she muttered.

In Chapter 13 Aquilon, still tracking the pattern's fate, exclaims: "This *R*-pentomino is a menace! I'm getting a headache! It just goes on and on." Gosper once said that to him the most impressive aspect of Conway's game is how it demonstrates the impossibility of predicting the outcome of processes that are rigidly determined by extremely simple rules of change. After Aquilon has learned about gliders and glider guns, she remarks: "If I were a pattern, I'd be very careful where I fired my gliders! That game plays a rough game!"

"It does," Cal replies. "As does all nature."

Much work has been done on variants of "Life": playing by other rules, and on other lattices such as triangular or hexagonal, and in dimensions higher than two. One-dimensional "Life" has also been explored—see the articles by Jonathan Millen and Munemi Miyamoto. "Life" has been investigated on wraparound fields that are cylinders and toruses, and even Moebius surfaces and Klein bottles. Some interesting results have emerged, but nothing compares with "Life" in the combination of richness of interesting forms with such simple transition rules. This is a tribute to Conway's intuition, and to the thoroughness with which he and his friends initially explored hundreds of alternate possibilities, including games with two or more sexes. Attempts have also been made to invent competitive games based on "Life," for two or more players, but so far without memorable results.

"Life" may have some practical uses. There have been attempts to apply it to socioeconomic systems, and a generalization of "Life" has been suggested as an explanation of why some nebulas have spiral arms (see the article by Kenneth Brecher). Arthur Appel and Arthur Stein, at IBM, found a way of applying rules similar to "Life's" in programs designed to identify the hidden edges in computer drawings of solid shapes.

I spoke earlier of the possibility that the universe is a vast cellular automaton, operated by the movements of ultimate particles (perhaps not yet discovered) according to unknown transition rules. Physicists are now searching for a GUT (Grand Unification Theory) that will bring together all the forces of nature into one unified theory based on a gauge structure. As physicist Claudio Rebbi explained in his article on "The Lattice Theory of Quark Confinement" (*Scientific American*, February 1983), a popular approach is to think of the gauge game as being played by particles on an abstract lattice of four-dimensional cubes—a sort of spacetime "Life." This suggestion was made in 1974 by Kenneth Wilson, and is now known as lattice gauge theory.

The game metaphor for GUT carries with it the implication

that the basic particles of the universe (pieces), the fundamental laws (transition rules), and spacetime (board) are not logical necessities. They are simply given. It is folly, as Hume and the positivists have taught us, to ask why they are what they are. Like chess players, physicists should accept the game and enjoy their (endless?) task of trying to guess how it is played, not waste energy speculating on why the game is designed the way it is. Now we are back to Leibniz and his stupendous vision of a transcendent Mind, contemplating all possible games, then choosing for our universe the Game that best suits the Mind's incomprehensible purposes.

BIBLIOGRAPHY

On cellular automata theory:

Theory of Self-Replicating Automata. John von Neumann. University of Illinois Press, 1966.

Computation: Finite and Infinite Machines. Marvin L. Minsky. Prentice-Hall, 1967.

Perceptrons. Marvin Minsky and Seymour Papert. MIT Press, 1969.

Cellular Automata. Edgar F. Codd. Academic Press, 1968.

Theories of Abstract Automata. Michael A. Arbib. Prentice-Hall, 1969.

Essays on Cellular Automata. Arthur W. Burks ed. University of Illinois Press, 1970.

On the Game of Life:

"Toward a Mathematical Definition of Life." Gregory J. Chaitin. Part 1, *ACM SICACT News,* Vol. 4, January 1970, pages 12–18; Part 2, *IBM Research Report RC 6919,* December 1977.

"The Game of Life: Is It Just a Game?" John Barry. *The London Times,* Sunday, June 13, 1971.

Lifeline, a newsletter on Life. Robert Wainwright ed. Issues 1 through 11, March 1971 through September 1973.

"The Game of Life." *Time,* January 21, 1974.

"Population Explosion: An Activity Lesson." Donald T. Piele. *Mathematics Teacher,* October 1974, pages 496–502.

"Life Games and Statistical Methods." M. Dresden and D. Wang. *Proceedings of the National Academy of Science,* Vol. 72, March 1975, pages 956–968.

"Life Line." Carl Helmers. *Byte,* Vol. 1, September 1975, pages 72–80.

"Lifeline." Carl Helmers. Part 2, *Byte,* October 1975, pages 34–42; Part 3, December 1975, pages 48–55; Part 4, January 1976, pages 32–41.

Ox. Piers Anthony. Avon, 1976. A novel involving Conway's Life and hexaflexagon theory.

"Statistical Mechanics of a Dynamical System Based on Conway's Game of Life." L. E. Schulman and P. E. Seiden. *IBM Research Report RC 6802,* October 1977.

"Self-Organization of Living Systems." Milan Zeleny. *International Journal of General Systems,* Vol. 4, 1977, pages 13–28.

"Life with Your Computer." Justin Millium, Judy Reardon, and Peter Smart. *Byte,* Vol. 3, December 1978, pages 45–50.

"Some Facts of Life." David J. Buckingham. *Byte,* Vol. 3, December 1978, pages 54–67.

"One-Dimensional Life." Jonathan K. Millen. *Byte,* Vol. 3, December 1978, pages 68–74.

"An Equilibrium State for a One-Dimensional Life Game." Munemi Miyamoto. *Journal of Mathematics of Kyoto University,* Vol. 19, 1979, pages 525–540.

"Life Algorithms." Mark D. Niemiec. *Byte,* Vol. 4, January 1979, pages 90–97.

"Life Can Be Easy." Randy Soderstrom. *Byte,* Vol. 4, April 1979, pages 166–169.

"The Game of Life." Howard A. Peelle. *Recreational Computing,* Vol. 7, May–June, 1979, pages 16–27.

"Spirals: Magnificent Mystery." Kenneth Brecher. *Science Digest,* Spring 1980, page 74ff.

"APL Makes Life Easy (and Vice Versa)." Selby Evans. *Byte,* Vol. 5, October 1980, pages 192–193.

"Life After Death." Pat Macaluso. *Byte,* Vol. 6, July 1981, pages 326–333.

"What is Life?" Chapter 25, *Winning Ways,* Vol. 2. Elwyn Berlekamp, John Conway, and Richard Guy. Academic Press, 1982.

INDEX OF NAMES